Vocational Award

WJEC

CONSTRUCTING THE BUILT ENVIRONMENT

LEVEL 1/2

Howard Davies

Published in 2019 by Illuminate Publishing Limited, an imprint of Hodder Education, an Hachette UK Company, Carmelite House, 50 Victoria Embankment, London EC4Y 0DZ

Orders: Please visit www.illuminatepublishing.com
or email sales@illuminatepublishing.com

British Library Cataloguing-in-Publication Data
A catalogue record for this book is available from the British Library

ISBN 978-1-912820-16-0

Printed by Ashford Colour Press, UK

11.22

The publisher's policy is to use papers that are natural, renewable and recyclable products made from wood grown in sustainable forests. The logging and manufacturing processes are expected to conform to the environmental regulations of the country of origin.

Every effort has been made to contact copyright holders of material reproduced in this book. If notified, the publishers will be pleased to rectify any errors or omissions at the earliest opportunity.

Editor: Dawn Booth
Text design and layout: Kamae Design
Cover design: Kamae Design
Cover photograph: Ant Clausen

You can download many of the forms throughout this book. Look out for the 'Downloadable' icon and visit www.illuminatepublishing.com/built_environment for access.

Contents

Author's Acknowledgements

Enormous thanks to my family, Lyn, Jac and Pip, for their encouragement and support during this adventure.

Heartfelt thanks to my faithful hounds Maisey (who sadly passed during the process) and Bummble who kept my feet warm.

Grateful thanks to all past and present colleagues, co-workers, designers, clients and wider stakeholders that continue to keep me busy and all those who engaged with the imagery and those who share the insight of the industry and its nuances. Thanks too, to Nick Chandler who took a chance on me and gave me my first job in the the industry.

Huge thanks to Allan Perry, Eve Thould and Dawn Booth for their professionalism.

Penultimate thanks to all learners, parents and teachers that engage with the built environment and its lifelong teachings.

Ultimately, thanks to Mother Nature for tolerating the industry and my Mother for her tolerant nature.

Consultant Editor: Steve Jones

Steve Jones has worked in the construction industry for over 40 years, starting his career as a bricklayer and progressing to Senior Project Manager, responsible for multi-million-pound projects. He now teaches at Coleg y Cymoedd. In 2009 he worked with a small team to set up the Construction qualifications for a major awarding body and has been a Chief Examiner since 2012.

Introduction

This student book has been designed and written to help guide you through the WJEC Level 1/2 Constructing the Built Environment course. It is divided into **three units** and then subdivided into sections that are identified with the learning outcomes (LOs), as used within the specification produced by the WJEC, to ensure understanding. These LOs have all been specially written in isolation so that you can focus on the LOs to demonstrate that you have the SKILLS, KNOWLEDGE, UNDERSTANDING and sometimes APPLICATION of characteristics of real work either through presentation of coursework or examinations.

This course provides genuine work-related learning by a plan, do and review approach, giving an opportunity to learn some fundamental and vital management, supervisory and tradesperson skills and outputs. Many of the LOs are transferrable in nature, providing you with a broad appreciation of the built environment and preparing you for advancement in further education, apprenticeships or careers within this vital sector.

The learning impacts on stakeholders in the industry, often referred to as the 'built environment', consisting of **a diverse group** of clients, designers, Engineers, tradespeople, specialist contractors and support staff, and supporting industries such as logistics, health and safety and information technology. The appreciation of this diversity and the collaborative approaches needed to help with good communication and outcomes is consistently referred to in all units of this book. The reliance on and relevance of our shared and limited elements of **time (programmes and defined durations)**, **health (of individuals and businesses alike)** and **resources (valuable and often difficult to obtain and control labour, plant and materials)** highlight and reference the relevance of **sustainability** within the built environment.

Within this book you will find

SKILLS These define the objective of the LO when it has been done well – for example, cooperating with your employer or those that work with or for you.

KNOWLEDGE This is the core of the information, facts and skills relevant to the LO – for example, what the Health and Safety at Work Act is.

UNDERSTANDING This is the awareness, comprehension and grasp of the factors that contribute to how you demonstrate your ability to deliver the LO – for example, our legal duties to cooperate with each other.

APPLICATION This is the action of putting something into operation – for example, completing a risk assessment or constructing a panel of brickwork.

You can download many of the forms throughout this book. Look out for the 'Downloadable' icon and visit https://www.illuminatepublishing.com/built_environment for access.

LINK

The Glossary of Key Terms is on pages 183–185.

KEY TERMS

These are short descriptions to highlight important words or narratives that help to summarise a concept. You will find a glossary of them at the end of the book.

REMEMBER

These are helpful and handy tips that help remind you how some concepts relate to the LO.

LINK

These show you where to read further in the book about a subject.

FACTS

These give important information that you need to remember.

 QR (quick response) codes take you quickly to the relevant webpage. The one on the left takes you to Illuminate's Constructing the Built Environment's dedicated website.

The Chartered Institute of Building website.

ISG/WJEC Construction YouTube channel.

How to succeed in this course

Basic software and access to the internet

Basic software and IT such as Microsoft Word, PowerPoint and Excel are essential to study this course, as you will need to practise completing some procedural forms, writing paperwork, completing paperwork and drafting very simple but helpful spreadsheets. Access to the internet is also required so that you can download free forms, templates and documents for this course from: https://www.illuminatepublishing.com/built_environment. By purchasing this book, you have access to all you need in terms of documents. The internet is a great source of knowledge, and the Chartered Institute of Building (https://www.ciob.org), trade organisations and other professional bodies have extensive learning resources. ISG plc, the author's employer, has teamed up with WJEC and produced a specific YouTube channel relating to this course that covers much of the unit content: https://www.youtube.com/channel/UCQSczLjBUv8ZWcf9mIuYv-Q/about.

'Why Study L1/2 Construction?'

A short film illustrating the components of the Level 1/2 Constructing the Built Environment course, called 'Why Study L1/2 Construction?' can be viewed at: https://www.youtube.com/watch?v=1rK5keCSaA0.

Access to a workshop

Everyone needs a safe and controlled environment to practise their skills, as this shows how you are progressing and the standards of your work. This area is needed to practise – and often fail, and if needs be fail again and again until it's right. **Everyone learns from their mistakes and tries not to let them happen again. If you do fail again, then revaluate your actions and move forward.**

Help

Everyone needs help and everyone also needs to help others. In many ways this is a theme of the course and the book and, as you will learn, it's part of law too.

Therefore, learners and teachers may find it helpful to **work alongside local construction businesses** and collaborate with them. This is not essential but is a meaningful and often enlightening way of understanding more about the industry. For example, ISG's vision is to be the world's most dynamic construction services company, delivering places that help people and businesses thrive. ISG is dedicated to changing the conversation about the construction industry: https://www.isgplc.com.

To deliver the places of tomorrow, the construction industry needs more people with the right skills to enter this exciting sector. ISG is passionate that we don't just talk about a skills gap; that we permanently close it.

That's why ISG joined forces with the Awarding Organisation, WJEC, to support the Constructing the Built Environment Level 1/2 course delivery as well as co-developing the Level 3 Applied Diploma in Professional Construction Practice (PCP), a qualification that is driving revolutionary change.

These new construction qualifications offer people the chance to become familiar with this great industry, introducing the standards, skills, techniques, technologies and ethos behind a career in construction.

ISG plc website.

The Level 3 PCP is set to become the gateway course for entry into careers in the world of property development, architecture, engineering, surveying and construction. Showcasing new and emerging technologies, the course is designed to provide students with knowledge and skills that the industry relies on.

To support the delivery of the Level 3 PCP, ISG has been working with an initial group of colleges across the UK to develop its Learning Alliance, allowing college tutors and ISG professionals to collaboratively plan the course delivery, maximising the opportunities for learners and the upskilling of industry professionals as trainers.

ISG's learning facilitators can provide useful links to industry bodies as well as exposure to live projects for practical experience. Together, this helps to ensure learners have a set of skills and knowledge that employers will truly value.

Course structure

WJEC details relating to the course.

Generally, you can start learning this course in Year 10 and in some cases Year 9. The course duration is over a two-year period and all the details relating to the specification, teaching guide, examiners reports and marking scheme can be viewed or downloaded from: https://www.eduqas.co.uk/qualifications/constructing-the-built-environment/index.

The unit content and assessment

WJEC Vocational Award in Constructing the Built Environment Level 1/2				
Unit number	Unit title	Guided learning hours	Assessment	Marks
1	Safety and security in construction	30	60-minute examination	60
2	Developing construction projects	60	Internal assessment by the centre	No examination
3	Planning construction projects	30	120-minute examination	60

The points available and their comparative value at GCSE level

	Points available per unit			
Unit number	Level 1	Level 2 Pass	Level 2 Merit	Level 2 Distinction
1	1	4	5	6
2	2	8	10	12
3	1	4	5	6

How can I manage and check the level of my understanding of work that I am expected to learn?

The Illuminate Constructing the Built Environment dedicated website, https://www.illuminatepublishing.com/built_environment.

The following table is a list of all the learning outcomes that you should know and be able to demonstrate by the time you submit your coursework or take your examinations, along with a brief summary of the assessment criteria. Keep track of your own perception of your progress by downloading it from the dedicated website or by copying it into your notebook and progressively ticking and tracking your strengths and weakness (you could also date your progress).

Unit 1 Safety and Security

LO	AC	Poor knowledge		Good knowledge		Great knowledge	
LO1 Know health and safety legal requirements for working in the construction industry	AC1.1 Responsibilities and safety legislation	AC1.1	☐	AC1.1	☐	AC1.1	☐
	AC1.2 Signs	AC1.2	☐	AC1.2	☐	AC1.2	☐
	AC1.3 Fire extinguishers	AC1.3	☐	AC1.3	☐	AC1.3	☐
	AC1.4 Role of HSE	AC1.4	☐	AC1.4	☐	AC1.4	☐
LO2 Understand risks to health and safety in different situations	AC2.1 Hazards	AC2.1	☐	AC2.1	☐	AC2.1	☐
	AC2.2 Effects	AC2.2	☐	AC2.2	☐	AC2.2	☐
	AC2.3 Risks	AC2.3	☐	AC2.3	☐	AC2.3	☐
LO3 Understand how to minimise risks to health and safety	AC3.1 Control measures	AC3.1	☐	AC3.1	☐	AC3.1	☐
	AC3.2 Situations	AC3.2	☐	AC3.2	☐	AC3.2	☐
LO4 Know how risks to security are minimised in construction	AC4.1 Security	AC4.1	☐	AC4.1	☐	AC4.1	☐
	AC4.2 Measures	AC4.2	☐	AC4.2	☐	AC4.2	☐

Unit 2 Developing Construction Projects

LO	AC	Poor knowledge		Good knowledge		Great knowledge	
LO1 Be able to interpret technical information	AC1.1 Interpret sources	AC1.1	☐	AC1.1	☐	AC1.1	☐
	AC1.2 Sources	AC1.2	☐	AC1.2	☐	AC1.2	☐
LO2 Know preparation requirements for construction tasks	AC2.1 Resources	AC2.1	☐	AC2.1	☐	AC2.1	☐
	AC2.2 Calculate	AC2.2	☐	AC2.2	☐	AC2.2	☐
	AC2.3 Success criteria	AC2.3	☐	AC2.3	☐	AC2.3	☐
	AC2.4 Prepare	AC2.4	☐	AC2.4	☐	AC2.4	☐
LO3 Be able to use construction processes in completion of construction tasks	AC3.1 Techniques	AC3.1	☐	AC3.1	☐	AC3.1	☐
	AC3.2 Health and safety	AC3.2	☐	AC3.2	☐	AC3.2	☐
	AC3.3 Evaluate	AC3.3	☐	AC3.3	☐	AC3.3	☐

Unit 3 Planning Construction Projects

LO	AC	Poor knowledge		Good knowledge		Great knowledge	
LO1 Know job roles involved in realising construction and built environment projects	AC1.1 Involve	AC1.1	☐	AC1.1	☐	AC1.1	☐
	AC1.2 Involve	AC1.2	☐	AC1.2	☐	AC1.2	☐
	AC1.3 Involve	AC1.3	☐	AC1.3	☐	AC1.3	☐
LO2 Understand how built environment development projects are realised	AC2.1 Processes	AC2.1	☐	AC2.1	☐	AC2.1	☐
	AC2.2 Factors	AC2.2	☐	AC2.2	☐	AC2.2	☐
	AC2.3 Sources	AC2.3	☐	AC2.3	☐	AC2.3	☐
	AC2.4 Resources	AC2.4	☐	AC2.4	☐	AC2.4	☐
LO3 Be able to plan built environment development projects	AC3.1 Processes	AC3.1	☐	AC3.1	☐	AC3.1	☐
	AC3.2 Processes	AC3.2	☐	AC3.2	☐	AC3.2	☐
	AC3.3 Tolerances	AC3.3	☐	AC3.3	☐	AC3.3	☐

Safety and Security in Construction

LO1 Know health and safety legal requirements for working in the construction industry

AC1.1 Summarise responsibilities of health and safety legislation

Any modern society acknowledges that its members have responsibilities, both to themselves and each other. These responsibilities are defined in our laws and acts of parliament that set out defined and accepted legal and moral obligations to one another. These laws and acts form the cornerstone of our society and help our industry to improve standards of the health, safety and welfare of its workers. The law related to this unit broadly defines the owners of these responsibilities as employers and employees. This learning outcome defines the importance of knowing the health and safety legal requirements of the construction industry. By having a knowledge of these responsibilities, we will be well placed to work compliantly.

KEY TERMS

Employer: A person or organisation who employs people under an employment contract.

Employee: Someone who works under an employment contract. A person may be an employee in employment law but have a different status for tax purposes. Employers must work out each employee's status in both employment law and tax law.

Compliant: An acceptable level of pre-agreed standards.

KNOWLEDGE / The HSWA

The 1974 Health and Safety at Work Act (HSWA) is an important act of parliament that is intended to be the primary legislation in England, Scotland and Wales.

The HSWA is intended to promote health and safety awareness, and effective standards of health and safety management are in place to promote, stimulate and encourage exacting standards of health and safety in the workplace.

This comprehensive act is intended to involve everyone in matters of health and safety. The groups can broadly be summarised as:

1 Management

2 Employees' representatives

3 Employees

4 Controllers of premises

5 Self-employed

6 Manufacturers of plant, equipment and materials.

The HSWA places **TWO** broad duties on employees:

1 **Section 7 (a) of the HSWA reminds us: To exercise reasonable care with the health and safety of themselves or others who may be affected by their acts or omission at work.**

2 **Section 7 (b) of the HSWA reminds us: To cooperate with the employer, as is necessary, to enable them (the employer) to comply (act in accordance with) with their legal duties in health and safety matters.**

KNOWLEDGE When a person becomes an employee they will enter into a contract with an employer, so both parties have responsibilities to each other in areas of the law:

1 **Contract law:** When two parties intend to form a legally binding agreement.

2 **Common law:** Law that has evolved (sometimes) over centuries where decisions are made in higher courts. This law looks to the past to find similar cases, called precedents, to follow the reasoning that may help to settle the matter fairly. This type of law deals with compensation if a party suffers loss, ill health or injury, and is also used to protect people from harm.

3 **Statute law:** This type of law is made by governments, the European Union (EU) and acts of parliament, and takes precedence over common law if there is a conflict between the two.

UNDERSTANDING Employees' responsibilities

Employees have very important duties under the Health and Safety at Work Act (HSWA), so they should always comply with the rules and instructions put in place by their employer whilst at work. If both parties follow this rule then compliance with the act is much more consistent than it would be if one or both parties failed to cooperate or communicate with the other.

Duties of the employee are to:

1 **Take reasonable care for the health and safety of themselves or others who may be affected by their acts or omissions.**

2 **Cooperate with their employer in all matters relating to health and safety.**

3 **Not intentionally or recklessly interfere with or misuse anything provided in the interests of health, safety and welfare.**

4 **Use anything provided by the employer in accordance with the instructions.**

5 **Report anything that is thought to be dangerous.**

'Health and Safety Tutorial', https://www.youtube.com/watch?v=eW7GBuaBuek.

APPLICATION Copy into your notebook and complete the table on page 12 with your own examples of positive and negative behaviours of an apprentice bricklayer on a residential construction site.

SCENARIO: It is winter, and the site is sloping with lots of operatives and vehicles working at the same time. The site is well managed and organised and nearing completion. Everyone is making their best efforts to complete the houses in time for Christmas. Refer to the employee duties list above to capture an example of each duty. Only the apprentice bricklayer is aware of the circumstances in all examples.

Employee duties	Scenario	Positive example	Negative example
1	A large puddle of frozen water has been created outside the canteen and reaches across both the footpath and road used for site pedestrian and vehicular traffic.		
2	The site manager has just posted a sign in the canteen that states: 'All operatives on site should inform their supervisor if they see any missing guard rails on the scaffold.' While working you noticed a carpenter tampering with a guardrail.		
3	The scaffold has just been erected on plot 101 and your supervisor has confirmed to you that it is safe to use. The access stairs are covered in ice.		
4	The site manager has just installed a hot and cold drinks machine in the canteen that all operatives can use free of charge. During the course of the day, you notice misuse of the machine.		
5	There is a scaffold board on the top level of scaffolding that appears to be about to break in half as it deflects and, as you have watched, another operative loads it with concrete blocks. It appears very weak and does not perform like the other boards.		

employee / employer

UNDERSTANDING Employers' responsibilities

Employers also have very important duties under the Health and Safety at Work Act (HSWA), so they should always comply with the rules and instructions put in place by their directors and sometimes themselves (often in the form of company policies) while employing workers. If both parties follow this rule, then compliance with the act is much more consistent than it would be if one or both parties failed to cooperate or communicate with the other.

Duties of the employer are to:

1 Assess and decide what could harm workers in their job and put precautions in place to prevent hazards causing issues. This is part of the risk assessment process.

2 Produce clear and easy to understand ways risks can be controlled in the workplace and who is responsible for doing this.

3 Consult and work with workers and their health and safety representatives to ensure that everyone is protected from harm when they are in work.

4 Provide workers with free training that is relevant to the job an employee is employed to do.

5 Provide workers with free equipment and personal protective equipment (PPE) and ensure workers maintain this equipment in serviceable condition.

6 Provide workers with adequate toilets, washing facilities and clean fresh water that is potable (drinking water).

KEY TERM

PPE: Personal protective equipment can include gloves, goggles and hard hats.

7 Provide workers with adequate and accessible first-aid facilities.

8 Report any major injuries and fatalities at work to the Health and Safety Executive's (HSE's) Incident Contact Centre: 0345 300 9923. Report other injuries, diseases and dangerous incidents online at https://www.hse.gov.uk.

9 Provide insurance that ensures workers are adequately covered by protective policies should they get hurt or become ill while working. These policies should be put on display, where they can be seen and read by all employees and visitors on site.

10 Employers must work with any other employers or contractors sharing the workplace or providing employees (such as agency workers), so that everyone's health and safety is protected.

'Information and Services', HSE

SKILLS / If employers and employees work collaboratively to plan and execute their work and maintain effective and productive communications, then it is likely that the workplace will be safer and more productive.

KEY TERM

Collaborate: Willingly and positively help to create or contribute to a project.

KNOWLEDGE / ## Legislation

The collective laws, known as legislation, that help both employers and employees to work compliantly, can be many and complex. For this unit we only need to understand what legislation is relevant and be able to summarise its relevance. This helps the industry to be safer by breaking down the high-risk activities in categories of work that historically have had high accident, incident and near miss statistics.

REMEMBER

Safety is everybody's responsibility.

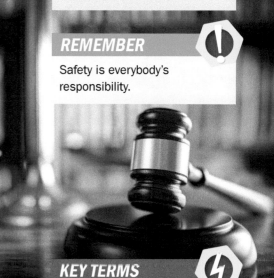

The eight categories of work that you need to understand are:

1 **Health and Safety at Work Act 1974**:
The primary piece of legislation that is intended to cover all occupational health and safety in the UK. Its aim is to protect all workers and wider stakeholders who might be affected by any work. The Health and Safety Executive, local authorities and some other organisations enforce the act.

KEY TERMS

Stakeholder: A member of a group with an interest in project.

Dutyholders: Those that plan or carry out the work.

2 **Reporting of Injuries, Diseases and Dangerous Occurrences Regulation 1995 (RIDDOR)**: If you are an employer, self-employed or in charge of premises and there is a serious accident then you must report it to the HSE within ten days. This includes:

- accidents resulting in the death of any person
- accidents resulting in specified injuries to workers
- non-fatal accidents requiring hospital treatment to non-workers
- dangerous occurrences.

If a worker is incapacitated for more than seven days, then the dutyholder must report this to the HSE within 15 days. Reporting can be done online or via telephone.

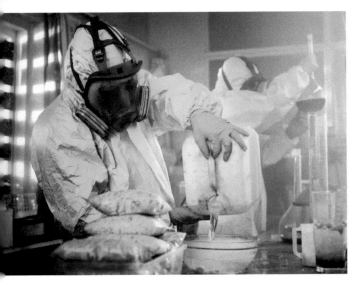

KEY TERMS

COSHH: The law that requires employers to control substances that are hazardous to health.

Monitor: To observe and conclude progress on a task or project.

3 **Control of Substances Hazardous to Health Regulation 2002 (COSHH):** Employers must and should take all precautions to reduce the exposure of their employees and anyone else to any substance that could potentially harm them or the wider environment, this includes:

- finding out what the health hazards are, such as vapour and contact with the skin
- deciding how to prevent harm to health (risk assessment), by qualified persons
- providing control measures to reduce harm to health, such as specialist safe systems of work and PPE
- making sure measures are used, such as specialised supervisors and methods of work
- keeping all control measures in good working order, such as keeping registers for their materials, well maintained plant and PPE
- providing information, instruction and training for employees and others, such as specialist courses often given by the chemical manufacturer or supplier
- providing monitoring and health surveillance in appropriate cases: private health companies and the National Health Service (NHS) offer this special service to contractors
- planning for emergencies, such as liaising with the local fire service, police and environmental agencies.

4 **Provision and Use of Work Equipment Regulations 1998 (PUWER):** Any business that owns, operates or controls tools or plant must ensure that this equipment is:

- suitable for the intended use, ensuring the appropriate specified tool is used for the job
- safe for use, maintained in a safe condition and inspected to ensure it is correctly installed and does not subsequently deteriorate, by ensuring staff are trained and equipment is checked regularly
- used only by people who have received adequate information, instruction and training, by keeping training up to date and relevant
- accompanied by suitable health and safety measures, such as protective devices and controls. These will normally include emergency stop devices, adequate means of isolation from sources of energy, clearly visible markings and warning devices. Manufacturers of equipment must ensure that these devices are available and easy to use
- used in accordance with specific requirements, for mobile work equipment and power presses, by not exceeding the capabilities of the equipment and selecting the right devices for the task.

5 **Manual Handling Operations Regulations 1992:**
If anyone needs to lift either a person, animal or
anything else (such as a concrete block) then the
employer must ensure that a simple hierarchy is used
to determine the safest way to lift prior to doing so:

- Avoid hazardous manual handling operations so
  far as is reasonably practicable, for example by
  finding an alternative method or device that will do
  the job.
- Assess any hazardous manual handling
  operations that cannot be avoided, by completing
  a special lifting risk assessment.
- Reduce the risk of injury so far as is reasonably
  practicable, such as dividing the load into small
  units, hiring specialist plant or devices, or
  seeking expert lifting advice from a construction
  professional.

6 **Personal Protective Equipment at Work Regulations
1992 (PPER):** Employers must provide free PPE to
their workforce, this will act as the first line of defence
for the employee. Nevertheless, PPE should be the
last resort as part of safe systems of work. This
means alternative methods and approaches to work
need to be used first, in order to reduce risk.

7 **Working at Height Regulations 2005:** Working at
height is currently the biggest killer in the construction
industry. Everyone in the industry should understand
this and understand that there are three simple steps
to avoid death or injury as a result:

- Avoid work at height if possible, by not placing
  stakeholders at risk.
- Where working at height cannot be avoided,
  prevent falls using either an existing place of work
  that is already safe or the right type of equipment,
  by fully utilising the latest plant and equipment
  readily available.
- Minimise the distance and consequences of a fall,
  by using the right type of equipment where the
  risk cannot be eliminated, such at safety netting,
  edge protection and scaffolding.

8 **The Control of Asbestos Regulations 2012:** If you
own, manage or have responsibility for a building
or if you are working on an existing building, these
regulations will apply to you. Fibres from asbestos
are potentially harmful if breathed in or digested by
the body, so these groups of stakeholders should be
aware that the regulations apply to them. There are
comprehensive measures that stakeholders must
take to comply with these onerous regulations.

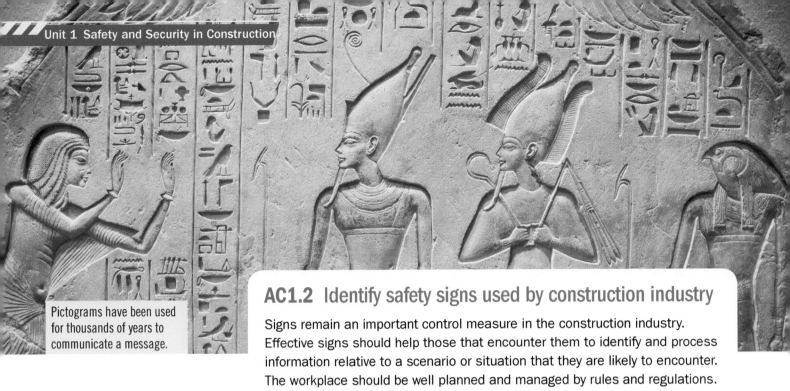

Pictograms have been used for thousands of years to communicate a message.

AC1.2 Identify safety signs used by construction industry

Signs remain an important control measure in the construction industry. Effective signs should help those that encounter them to identify and process information relative to a scenario or situation that they are likely to encounter. The workplace should be well planned and managed by rules and regulations. Consistency of signs that people encounter is very important so the Health and Safety (Safety Signs and Signals) Regulations 1996, which informs us what signs should look like. The regulations generally apply to employers, employees, dutyholders and others (i.e. members of the public). Sometimes, this broad group is referred to as 'stakeholders'.

REMEMBER

Signs should be simple and easy to understand.

KEY TERM

Identify: Recognise, distinguish and establish what something is.

FACT

Construction site traffic signs should meet the latest requirements of the (road) Traffic Sign Regulations, so are not covered by the syllabus of this qualification.

REMEMBER

Colour code and shape help us to understand types of signs from a long distance or when visibility is poor.

SKILLS // Safety signs

It is vital that stakeholders can identify the intended meaning of a sign and, to make this possible, signs have varying colours, shapes and content to help make prioritising and reading them easier and more informative. If all signs had the same format, then it is likely that many stakeholders would soon ignore them as they would become too familiar. So, construction signs in Great Britain and the EU should have a consistent format. This is so stakeholders who live, work and travel in Europe have a standard set of recognisable signs to help keep them safe from harm and aid them to travel safely and efficiently.

KNOWLEDGE // There is a limited assortment of colours and shapes to signs in the construction industry, these are:

RED circle = Prohibition sign (must not do!)

YELLOW or AMBER triangle = Warning (warn of a hazard or danger)

BLUE circle = Mandatory sign (must do!)

GREEN rectangle = Emergency escape and first aid (information on routes and locations)

RED rectangle = Fire-fighting sign (location of fire-fighting equipment)

All signs have specific coloured pictograms. The red prohibition sign pictogram has a white background to help it stand out. It also has a red diagonal line struck through it.

The category of construction safety signs can be understood by identifying the colour, shape and content of the relative sign. Take time to recognise and understand the pictogram within the sign. Sometimes there may be more than one pictogram within the sign, to highlight the potential cause and effect of the hazard, or sometimes the hazard and control measure. Look at the translated signs below.

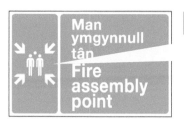

DO NOT

 No naked flames (naked flames are prohibited)

 No access for pedestrians (pedestrians are prohibited from accessing)

MUST DO

 Head protection must be worn

 Foot protection must be worn

WARNING

 Risk of fire

 General warning sign (often accompanied by another sign informing you of the associated hazard)

EMERGENCY

 First aid (often accompanied with a route arrow or another pictogram)

FIRE-FIGHTING

 Fire alarm call point (see how this sign has the hazard and control measure on it)

UNDERSTANDING Signs are another form of control measure, so should be located where they will be seen by as many stakeholders as possible. Some signs may be used with other types of communication tools such as:

- hand signals and verbal communication
- illuminated signs with verbal communication
- illuminated signs with acoustic signals.

KEY TERM

Site induction: The formal presentation given to all new visitors and members of the workforce when they arrive on site and always before they enter the site on a visit or to work. A record of this safety critical presentation is kept to demonstrate that all stakeholders are aware of project-specific risks, hazards and objectives, as well as rules and restrictions.

REMEMBER

Shape ✓

Colour ✓

Pictogram ✓

APPLICATION In your notebook, design signs for the following, without the use of words:

1 No running on site

2 No dancing on site

3 Sunglasses and hoodies must be worn

4 Danger, giant hot dog

5 Noses must be pierced

6 Fast food this way

7 Emergency phone-charge point.

Using the image below, complete the missing words for the signs 1 to 8. Research the meanings of the other signs on the page, 1–44, and record you findings in your notebook.

For your Health & Safety:

1 No _____

2 __ smoking

3 __ _____

4 Wear a ____ hat

5 Wear safety _____

6 ____ high-visibility vest

7 Wear ___ protection

8 ____ protective gloves

Attend a Health & Safety induction before starting work

All visitors must be inducted and accompanied

See Health & Safety plan for evacuation procedure and muster point

Only pre-booked deliveries will be accepted

AC1.3 Identify fire extinguishers used in different situations

The **risk** of fire occurring within the construction industry is often highly likely unless control measures are put in place. The outbreak of fire can often appear to be spontaneous and frequently spreads rapidly unless an intervention is made as early as possible but only if it is safe to do so. The outbreak of fire within the construction industry presents a risk to workers' lives, health and reputations. The resulting damage caused by fires can lead to considerable commercial loss and litigation. The consequences of fires on construction sites often lead to delays in completing projects, resulting in further commercial loss and litigation, which compounds an already potentially devastating, unwanted and often preventable dangerous occurrence.

KEY TERM

Risk: The likelihood that a person may be harmed if they are exposed to a hazard.

SKILLS / Knowing about potential fire hazards and having the skills to safely and effectively check, report, raise the alarm and/or intervene are vital and potentially lifesaving skills to possess. Interventions such as these will likely raise awareness and mitigate potentially deadly and costly fire spread.

REMEMBER

Fires don't just happen, they are caused!

KNOWLEDGE / **What is fire?**

Fire is a chemical reaction between three factors: oxygen (usually within the atmosphere), a type of fuel (such as wood or paper) and ignition (such as a spark). This relationship is often referred to as the 'fire triangle'. If any one or more of the three is removed or not present, then the likelihood of fire is reduced.

There are distinct types of fire, which are placed into categories and referred to as 'class of fire'. So, if any class of fire occurs on a construction project or while 'at work', there is often a certain type of reactive firefighting equipment available to extinguish it or slow its spread to other areas. In turn this provides valuable time for workers to raise the alarm, put the fire 'out', allow workers to escape to a safe place and/or limit the spread of fire.

KEY TERM

Likelihood: The chance something will happen or the probability of something becoming a reality.

FACT

Do not make the mistake that all fires are the same, they are not!

UNDERSTANDING / The most practical reactive control measure to use in the event of a fire on site is a manufactured product commonly known as a 'fire extinguisher'. There are distinct types of fire extinguishers designed and manufactured to help fight the corresponding class of fire. It is important to understand which type of extinguisher is the most effective and appropriate to use, so let's look at them.

Pretend it's a fruit smoothie!	Class (type) of fire	Extinguishing agents	Identification colour
Cream	Liquid fires	Foam	CREAM
Strawberry	Wood, paper, textile and all solid fuel fires	Water	RED
Blueberry	Electrical fires	Powder	BLUE
Blackberry	Liquid and electrical fires	Carbon dioxide	BLACK
Banana	Wood, paper, cooking oil, textiles and solid material fires	Wet chemical	YELLOW

Fire blankets are the other type of commonly used control measure to effectively 'smother' a fire, by reducing the amount of air around the ignited fuel. These are intended to be used on fires that are contained in vessels such as saucepans, pots and containers. They are also effective when tackling clothing that may be alight while being worn.

All fire extinguishers in the UK are red in colour and should have an identification extinguishing agent colour. This is a carbon dioxide fire extinguisher in use.

KNOWLEDGE / **Fire risk assessments**

We now understand that hazards associated with the use and storage of materials in the construction industry are likely to be greater if we don't manage the workplace adequately. To ensure this management is as thorough as possible, a fire risk assessment is conducted by the manager or supervisor and is completed as shown in the sample on the following page.

1 Identify and determine hazards: the types of materials that are combustible.

2 Identify the people at risk: specific groups of workers and people.

3 Evaluate and act: consider steps 1 and 2 and remove/reduce risks.

A fire risk assessment helps the manager or supervisor to ...

4 Record, plan and train: make sure you have a record of what and where the hazards are and that there's an emergency plan in place and you have trained operatives that can carry out procedures.

5 Review: regularly review your risk assessment and update it to ensure it captures the current scenarios and conditions of site.

Fire Risk Assessment

Project Name:	New house	Completed by:	HD		Date:	23 March 2020	Revision:		A	Page: 1 of 1	
Location:	Within site storage container	People affected/at risk of being harmed: Site Operatives, Site Management, Visitors and 3rd Parties									

No.	Hazards	Possible affects/harm	Pre-control risk rating			Required controls	Post-control risk rating		
			High	Medium	Low		High	Medium	Low
1	Dry wood store		✓			Keep store locked, secure and away from designated smoking area. Place danger and warning signs outside and within store. Place a water fire extinguisher at nearest fire point. The Fire Marshall will check the store is clean and tidy at the end of every day.			✓
2									

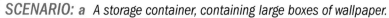

APPLICATION

1 Complete the blank fire risk assessment, which can be downloaded from the dedicated website and is shown on the next page, based on the following scenarios:

SCENARIO: a A storage container, containing large boxes of wallpaper.

b A busy site canteen that cooks food.

c A small room on the project that is nearing completion, where welding is ongoing.

The Illuminate Constructing the Built Environment dedicated website, https://www.illuminatepublishing.com/built_environment.

UNDERSTANDING

REMEMBER

If you need some inspiration to help you recall the colour and purpose of fire extinguishers try to imagine an easily identifiable food that is the same colour as that of the code of the extinguisher (extinguisher agent) in context of the 'class of fire'.

Yellow bananas and cooking oil.

Downloadable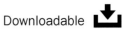

Fire Risk Assessment

Project Name:		Completed by:		Date:		Revision:	A	Page: 1 of 1	
Location:		People affected/at risk of being harmed:							

No.	Hazards	Possible affects/harm	Pre-control risk rating			Required controls	Post-control risk rating		
			High	Medium	Low		High	Medium	Low
1									
2									
3									
4									
5									
6									
7									
8									
9									
10									

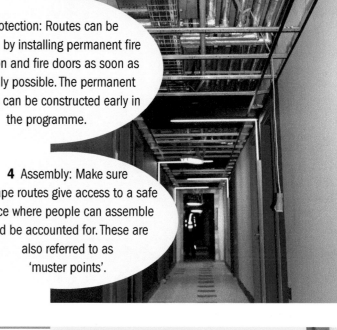

Means of escape

It is essential that construction sites have adequate and defined routes for escaping from the workplace in the event of a fire (and/or another emergency). The provision of these specifically planned and maintained routes that lead the workforce away from danger are a fundamental control measure on any construction site.

Key aspects to providing safe means of escape on construction sites are shown on the following mind map.

2 Alternatives: Well-separated alternative ways to ground level should be provided where possible. This could be in the form of additional temporary stair systems.

3 Protection: Routes can be protected by installing permanent fire separation and fire doors as soon as practically possible. The permanent partitions can be constructed early in the programme.

1 Routes: The routes in the risk assessment should determine the escape routes required – these must be kept available and unobstructed.

Means of escape

4 Assembly: Make sure escape routes give access to a safe place where people can assemble and be accounted for. These are also referred to as 'muster points'.

5 Signs: Escape routes need good lighting and signs will be needed to identify them. Emergency lighting may be required for larger and more complex projects.

Means of giving a warning

It is vital that a system should be installed to alert everybody on site of a fire, should one occur. This may be some temporary or permanent mains-operated fire alarm, which should be tested regularly by a competent person. This alert/alarm could be in the form of a:

✓ Klaxon   ✓ Air horn   ✓ Whistle   ✓ Electronic

The warning needs to be distinctive, audible above other noise and recognisable by everyone.

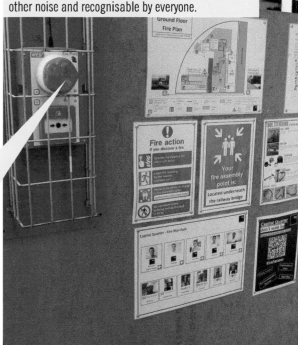

PUSH ME

Means of fire-fighting

Trained and nominated people should be appointed to use fire extinguishers. Everybody on site should know who these people are.

Fire extinguishers should be located at identified fire points (as shown below) around the site. The extinguishers should be appropriate to the nature of the potential fire:

- wood, paper and cloth – water extinguisher
- flammable liquids – dry powder or foam extinguisher
- electrical – carbon dioxide (CO_2) extinguisher.

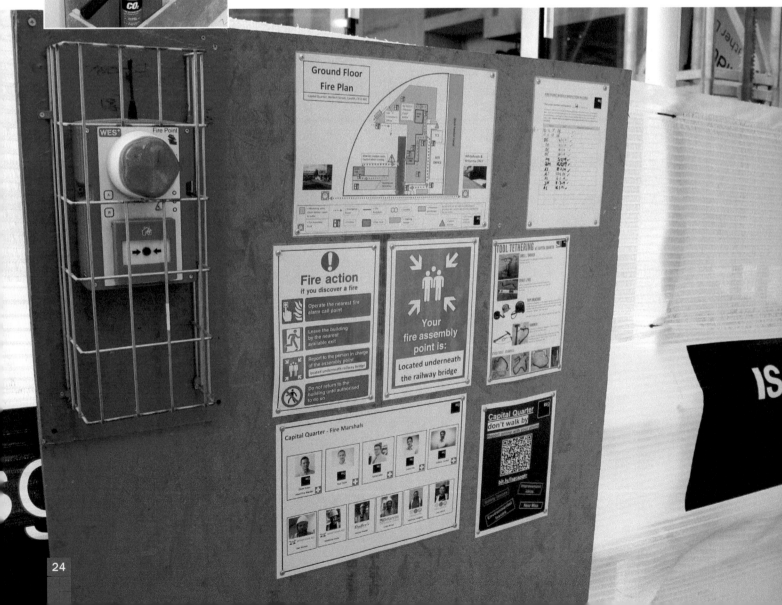

AC1.4 Describe role of the Health and Safety Executive

The Health and Safety Executive (HSE) is a centrally funded government and locally managed organisation. Its officers are a diverse group of professionals who all play a vital role in keeping the UK's workforce healthy and safe. Its mission is to prevent death, injury and ill health in the UK's workplaces. The HSE is a statutory body established by the Health and Safety at Work Act 1974 (HSWA). In Northern Ireland it is referred to as the Health and Safety Executive Northern Ireland (HSENI) and has a different logo.

CONTROLLING RISK TOGETHER

HSE ARE NOW IN YOUR AREA

SKILLS // The HSE

Everyone in the UK, regardless of occupation, should understand what the HSE is, and respect and follow its advice. In the construction industry this advice can be in the form of **proactive** distribution (controlling the distribution so the material is available when needed) of helpful information, best practice and suggesting guidelines, whatever construction discipline you represent or work within. The HSE also issues **reactive** legally binding advice in the form of 'notices', if they consider there to be a 'breach' of legislation or that improvements to working practices are required.

ARE YOU COMPLIANT?

Find out more information: www.hse.gov.uk/construction

Join the conversation at: SaferSites

KNOWLEDGE // The Secretary of State (SoS) has principal responsibility for HSE. The Department of Work and Pensions (DWP) minister with responsibility for health and safety will account for HSE's business in Parliament, including its use of resources. The HSE Board is responsible to the relevant ministers for the administration of the 1974 Act.

HSE's main statutory duties are to:

1 Propose and set standards for health and safety performance, including submitting proposals to the relevant SoS for health and safety regulations and codes of practice.

2 Secure compliance with these standards, including making appropriate arrangements for enforcement.

3 Make arrangements, as it considers appropriate, for the carrying out of research, and the publication of the results of research, as well as encouraging research by others.

4 Make arrangements, as it considers appropriate, for the provision of an information and advisory service, ensuring relevant groups are kept informed of and adequately advised on matters related to health and safety.

5 Provide a Minister of the Crown, on request, with information and expert advice.

> **FACT**
>
> Members of the HSE wear yellow hard hats with HSE's famous logo on the front.

> Our mission is to prevent death, injury and ill health in the UK's workplaces.

> **FACT**
>
> The HSE 'police' all sectors of industry in the UK and have powers. It uses 'section 20' to enforce the law. This section of the act allows the HSE inspector to enter any premise where they have any reasonable doubts that the law may be compromised. They can even have a police escort to do so and survey the situation by taking photographs, factual surveys and interviewing witnesses.

The HSE deploys specially trained, qualified and experienced 'inspectors' and 'officers' to visit construction projects. The inspectors are based in a network of local and regional offices so that they are familiar with their local environment. This makes them very well informed and effective.

APPLICATION // Find out how many powers the HSE has.

UNDERSTANDING // When things go wrong!

If an HSE inspector visits a site and considers the works to be unsatisfactory, they can issue the management with warning notices, which is how the HSE enforces health and safety law.

Serving **notices** on dutyholders. Formal notices that can force you to stop (**prohibit**) and/or **improve** how you are working, such as creating too much noise, dust or vibration.

Providing information and advice face-to-face or in writing, such as data sheets on the best methods to approach a certain activity such as traffic management of a narrow busy road.

Withdrawing approvals. If you don't have approval to work, then you don't work!

Prosecution. This is the most severe form of legal enforcement, which could result in a fine or even a prison sentence.

The HSE can ...

Varying licences, conditions or exemptions. This could change your method, timings or general approach to the 'works'.

Issuing simple cautions. The inspector may have seen a minor breach of the regulations such as a worker not wearing appropriate PPE.

APPLICATION // Investigate the largest fine and the longest prison sentence awarded for breach of health and safety law in the UK. Record these in your notebook.

UNDERSTANDING Prevention through advice and support. The HSE provides proactive, helpful advice on many aspects of health and safety to the construction industry. It provides detailed and informative publications and digital tools that are intended to prevent accidents, incidents and near misses in the construction and 'all other' industries. According to HSE's website, it provides:

Advice and guidance on the law. We provide targeted advice, information and guidance to help **dutyholders** *comply* **with health and safety law** *in* **a sensible and proportionate** *way.*

We use several methods including:

simple, practical advice to help small businesses **understand** *the risks and what they need to do about them.* (http://www.hse.gov.uk/enforce/advice-information-guidance.htm)

APPLICATION Visit the HSE's 'Advice and Guidance on the Law' page: http://www.hse.gov.uk/enforce/advice-information-guidance.htm.

Spend some time navigating the pages of the website. Then why not sign up to the digital alerts that the HSE offers to keep you up to date with construction-related fines and sentences that have been awarded? Make notes in your notebook of the circumstances of the relative incidents. It will soon become apparent how many H&S issues arise across all industries in the UK.

APPLICATION The HSE collects, studies, shares and reports on statistical data relating to all health matters in the UK. This allows government and industry to target and deliver continual improvement on how we approach safety when delivering the built environment.

Read online the HSE's 'Construction Statistics in Great Britain, 2018' at: http://www.hse.gov.uk/statistics/industry/construction.pdf.

Write your answers to the following in your notebook:

1 Name the activity that caused the most deaths.

2 Name two work-related/ill health illnesses.

3 What percentage of the UK workforce is employed in construction?

4 How many working days are lost due to accidents at work?

5 Give two examples of prohibition notices and improvement notices relevant to construction.

'Advice and Guidance on the Law', HSE.

'Construction Statistics in Great Britain, 2018', HSE.

LO2 Understand risks to health and safety in different situations

AC2.1 Identify hazards to health and safety in different situations

Considering that a **hazard** is anything with the '**potential to cause harm**' and construction sites change rapidly due to the complexity of what is under construction, along with how they are managed and supervised, then it's fair to say that the construction industry **both on and off site potentially can be harmful**.

KNOWLEDGE Every year workers in the industry are **killed** and **injured** both **on** and **off** site. The reasons for such accidents, incidents and near misses have many causes but the circumstances can broadly be categorised into the groups shown in the following table, which shows the greatest number of incidents ranked in order of frequency.

FACT

Thirty-eight construction workers died in 2018.

FACT

Fifty-eight thousand construction workers suffered injuries in 2018.

Deaths in construction (UK 2018)	Non-fatal injuries in construction (UK 2018)
1 Falling from height	1 Slips, trips and falls
2 Trapped by something collapsing	2 Injuries from handling, lifting or carrying
3 Struck by an object	3 Falls from height
4 Struck by a moving vehicle	4 Struck by something moving, including a flying/falling or moving object
5 Contact with electricity	

(Source of table and fact boxes on left: RIDDOR, cited in HSE (2018, 31 October) 'Construction Statistics in Great Britain, 2018', http://www.hse.gov.uk/statistics/industry/construction.pdf)

The situations of both death and non-fatal injuries are similar to previous years.

By comparing the actual deaths and non-fatal injuries statistics in the table above we can determine what is likely to kill and/or injure people both on and off site. This means we can also proactively target these same broad groups to **mitigate such potentially devastating and life-changing events from ever happening**.

SKILLS By taking **positive and proactive action** now because we are aware of the types of **hazards** and the actual **situation** that cause deaths and non-fatal injuries, it's also fair to say that it is likely that an accident, incident or near-miss **will not occur** when carrying out a related activity in the future. This is how **risk** is mitigated. We can ensure that we incorporate **control measures** into how we **plan, manage and execute** the same construction activities in all these situations in the future.

Falling from height is the biggest cause of death in the construction industry, so all precautions should be taken.

KNOWLEDGE / **Hazards when forming substructures** (such as constructing foundations, drainage and services) are shown in the following table.

Likely to cause deaths	Likely to cause non-fatal injuries
1 Fall into a deep excavation	1 Walking on uneven and/or rough terrain to get to and from the workplace
2 Crushed by the collapsed sides of a deep excavation that has no earthworks support (shoring)	2 Failing to have adequate manual handling training, lifting equipment or operators to minimise the lifting and carrying of heavy materials (of any weight)
3 Hit on the head by a large boulder dislodged by rain fall when working at the base of a trench	3 Falling off a ladder while climbing down into an excavation
4 Sharing the same space as moving vehicles without adequate segregation in place	4 Not wearing eye protection when working within a dry/dusty trench on a very windy day
5 Not checking the survey drawings for the location of existing live services prior to excavating the site	

KNOWLEDGE / **Hazards when forming superstructures** (such as steel erection, brick/block laying, erecting scaffolding and timber frames) are shown in the following table.

Likely to cause deaths	Likely to cause non-fatal injuries
1 Fall from a scaffold/trestle at any height	1 Breaking an ankle in a void within a concrete floor
2 Crushed by an unsupported blockwork wall during high winds	2 Hurting your back while bending over to lift a load
3 Hit on the head (while you are working) by a falling scaffolding spanner dropped by a worker above you	3 Falling 600mm from a trestle
4 Hit by a mobile elevated platform	4 Hit on the head (while you are working) by a falling bolt dropped by a worker above you
5 Cutting through a live power cable while chasing a wall	

KNOWLEDGE / **Off-site hazards** (such as joinery workshop, offices and travelling between sites) are shown in the following table.

Situation	Likely to cause deaths	Likely to cause non-fatal injuries
Workshop	• Electrocution from a faulty power supply • Crushed by a forklift truck that was reversing in a busy warehouse	• Chest infection from breathing wood dust • Loss of sight in one eye from a foreign body entering it • Severed fingers from a faulty mechanical saw blade
Office	• Falling from a chair used as access while changing a lightbulb • Heart attack while working alone late at night	• Work-related upper limb disorder from heavy lifting • Back injury from poor posture over an extended period while using a computer monitor
Travelling	• Hit by a vehicle while exiting from a car on a busy road • Crushed while using a mobile telephone in a busy factory car park • Death caused by a speeding offence	• Road traffic accident caused through falling asleep while driving long distances at night • Road traffic accident caused by a speeding offence

APPLICATION / Look at the images on this page of actual safety failings and consider the health implications of the those who could be directly and indirectly affected. In your notebook write down the potential causes of:

1 Death

2 Non-fatal injury.

AC2.2 Describe potential effects of hazards in different situations

Most people in their everyday lives, going about their everyday business, have three things in common. These are precious and difficult to see and touch but easy to understand. Regardless of age, sex, race, gender, religion and many other things that define us as human beings, they remain the same, these are:

- Health
- Time
- Resources

It is vital to respect your own 'three things' as much as everybody else's, and those of every living thing also. That's why the workplace has laws, regulations, rules, processes and procedures. These help with the governance of the workplace.

SKILLS Accidents, incidents and near misses rarely just happen, **they are caused**. When these do occur, let's focus on the effects upon our 'three things'.

APPLICATION Most of us have been hurt in some way when we have had an accident. Complete and record in your notebook the 'empathy' table below with the details of any accident you have experienced or witnessed (keep this private).

What happened?
Cause?
Length of effect?
What did it prevent or hinder you doing?
Pain scale 1–10 (10 = very painful)
How did you feel?

By undertaking this exercise, you have a basis of comparison that allows you to empathise and expand your understanding of the potential power hazards have if they develop into risks and go unchecked in the workplace.

KNOWLEDGE There are direct effects of hazards in the workplace that present themselves when hazards go unchecked, these are:

The **physical** effects on the people involved such as personal injury, sickness and illness. Living and coping with long-term injuries can have additional consequences to your health. Musculoskeletal disorders can prevent you from being active and impair and/or prevent you from doing a job you worked hard to get. Respiratory disease can shorten your working life and can be fatal.

The **psychological** effects on the people involved such as workers, managers, the public, neighbours, visitors, tourists and commuters can cause stress-related conditions and depression and have long-lasting, sometimes irreversible, consequences on mental health. Anxiety, shame and despair all play a part in eroding a person's wellbeing.

KEY TERM

Contaminated: Exposed to pollution or chemicals/substances.

The **financial** effects of loss of earnings due to having to take time off work to recover, paying fines imposed for breaches of legislation, or payment for compensation following an accident involving a member of the public. Having to clean-up contaminated areas is a specialist occupation and very costly. Additional outgoings for therapy and associated medical costs can quickly build-up and increase over time.

The **environmental** effects of hazards can cause pollution and contaminate the air with noise and dust. Watercourses such as streams and rivers can become contaminated for lengthy periods of time by chemical spills, while town centres may become uncomfortable and feel confined due to excess dust, vibration and poor management and supervision.

'Bricklaying Tutorial', https://www.youtube.com/watch?v=7bWI3m19AYU.

KNOWLEDGE The hazard spectrum is a comprehensive tool we can use to capture the potential effects of a hazard. Think of a pebble being thrown into a 'still' pond and causing a ripple, as the ripple expands the wider the impact on the surface.

SCENARIO: Megan, a bricklayer, is using a disc cutter in a city centre without utilising adequate PPE or the dust suppression that she was given and failed to follow the method statement or risk assessment that she was briefed on during her induction.

Time	START

09:30 — Self

Megan receives dust in her eyes and inhales substantial amounts of silica dust. Megan must stop work and go to hospital immediately.

09:35 — Other workers

Fourteen of Megan's co-workers are directly contaminated by silica dust and must stop work. Freshly painted windows are ruined and the HSE inspectors have just walked onto the site to carry out an audit. They have seen what has happened.

09:40 — Employer

Megan's supervisor receives a written warning from the site manager. The HSE issues a prohibition notice to the site manager, so the site must be closed until further notice. The site manager must call head office and he is summoned to line management for a disciplinary hearing.

12:00 — Local community

The local press has taken images of the incident and hear about the HSE issuing a prohibition notice to the site. The story is published on social media. The local councillor gets 12 complaints from businesses and pedestrians. The council decide to remove the contractor from its approved list of contractors for one year to demonstrate to the local tax payers that they are listening to the concerns of the public.

15:00 — Environment

It's a lovely sunny, dry day and the local businesses demand the contractor cleans down the facades of their new offices and complain that the silica dust has scratched their new glazing. The business decides to combine forces and employ a local solicitor who is a member of their group and launch a court action against the contractor.

16:30 — Users

The town centre pedestrians, businesses and local authority are wary of the contractor and look for other opportunities and instances to protest about how it works. Morale on site is now poor and relationships, reputations and quality deteriorate. Megan is off work and is waiting for an eye operation and further tests on her lungs as she doesn't feel well. Her co-workers must have long-term health surveillance since the incident.

KNOWLEDGE // The incident discussed above demonstrates how rapidly the actions of one individual can impact on lives, reputations of a business and the image of the construction industry. **Safety is everybody's responsibly** and the potential effects of hazards should never be underestimated.

APPLICATION Copy into your notebook and complete the hazard spectrum table below, based on the following scenario:

SCENARIO: It's Monday afternoon and a large and reputable electricity provider, which has a contractor working at the entry to a primary school, is excavating a trench for a new supply cable. A worker has spilt petrol while filling up a small generator and a passing taxi driver has ignited the fuel by throwing out a lit cigarette from his window. There has been a fire, which has been extinguished but sadly one operative is badly injured with severe burns. There is fuel on the road and it is 15:30 hrs (hours) so parents are starting to pick-up their children from school. The emergency services are investigating, and the casualty has just left site in an ambulance. There is a chaotic scene outside the school as a local news camera crew and reporters arrive to report on the incident. The tailback of traffic is now half a mile long. Businesses in the area are experiencing a large decrease in sales.

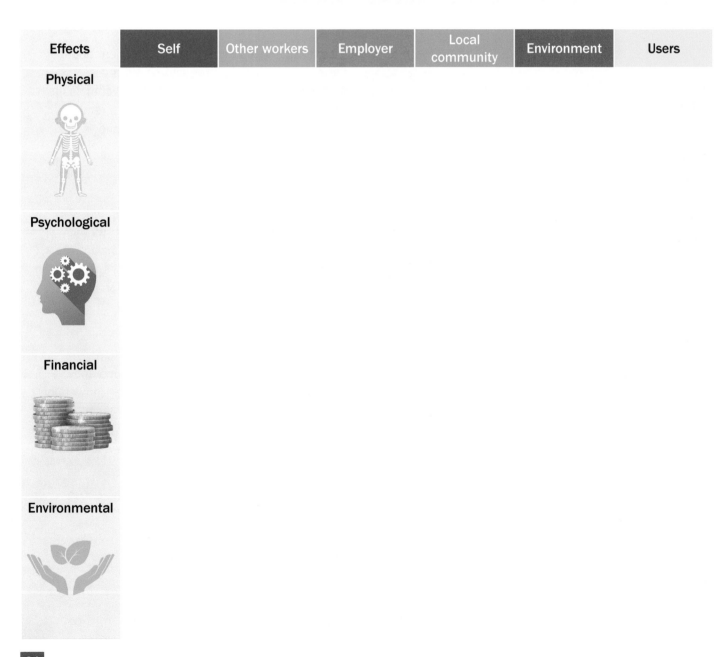

Effects	Self	Other workers	Employer	Local community	Environment	Users
Physical						
Psychological						
Financial						
Environmental						

AC2.3 Explain the risk of harm in different situations

All those that work within the construction industry are exposed to levels of risk. There is an infinite number of circumstances and reasons that influence the level of risk exposure, so it is everybody's responsibility to make the workplace as safe as possible. **Everybody** who works within the industry should be familiar with their own task-specific risk assessment. Some tasks are more hazardous than others so that's why assessment must be specific to the task or circumstance.

SKILLS You need to have the skill to calculate, record and comprehend the likelihood and severity of a hazard or potentially hazardous situation to ensure control measures are agreed and put in place prior to commencement. By doing so, you take control of situations and therefore are likely to work more safely, be more productive, and live longer and safer lives as a result. Understanding the resulting **measured risk** is therefore an approved, accepted and effective management tool that is easy to use and a **demonstrable way of complying with the law**.

KEY TERM

Severity: The degree of harm that could occur or a measure of how bad an injury could become.

KNOWLEDGE If you are to plan, manage and execute work safely, you should factor in the following elements prior to producing a risk assessment. These are:

✓ **Check manufacturer's instructions** which often have associated data sheets that may be relevant to things such as mechanical moving parts, limitations of the machine or device and chemical data sheets (known as COSHH sheets).

✓ **Interrogate your ill-health records** as these can often indicate areas where there has been a failure or weakness historically such as lack or supervision or poor training of staff.

✓ **Consider non-routine activities** such as breakdowns of machinery, emergency shutdowns or new working routines/methods that are required.

✓ **Long-term implications** of working with certain types of liquids, chemicals or materials that cause excessive dust or noise. Think what could be done to minimise exposure to these hazards.

✓ **Use the HSE (HSENI in Northern Ireland)** and its detailed and extensive knowledge-sharing website and always coordinate the work with qualified health and safety experts and construction professionals.

LINK

For more on the HSE see pages 25–27.

FACT

When likelihood is high, and severity is great, then death or severe injury will likely occur.

When likelihood is low, and severity is low, then a minor injury is possible and could occur.

FACT

When likelihood is high, and severity is great, then death or severe injury will likely occur.

When likelihood is low, and severity is low, then a minor injury is possible and could occur.

REMEMBER

Please be 'reasonably practicable'.

KEY TERMS

Eliminate: Get rid of entirely.

Mitigate: Get rid of as much as possible.

REMEMBER

The risk of harm is reduced because both the severity and likelihood of noise has been reduced. This risk assessment proves it.

Risk of harm is an accepted part of everyday life and you are **not expected to have no risks at all**. You should **identify** what the main **risks** are and understand what should be done to **manage them responsibly**.

If ever likelihood and severity are considered moderate, then death, injury or injury should be considered **likely**. A risk assessment helps you to easily and demonstrably understand each hazard's severity and probability of occurring.

KNOWLEDGE // ## How risk is measured

Referring to the example extract of a **risk assessment** on page 37, you can see that the busy Joinery Shop this refers to has a **potential noise hazard**. By studying this **risk assessment** you can see that the joiner has identified the type of **hazard** and who might be affected by the noisy machinery. Next, you can identify what measures the joiner already has in place to eliminate or mitigate the effects of the hazard on the workforce.

✓ The joiner has identified that improvements would be reasonably practicable to do by installing anti-vibration mouldings on certain machines to reduce impact and airborne noise within the workplace.

✓ The joiner has then identified that when the company orders some new machinery, they will ensure that the manufacturer will provide noise-reduction features to help reduce the transmission of airborne and impact noise.

✓ The joiner then identifies the person who is responsible to consider the modifications to the machines and order the new ones.

✓ The joiner demonstrates that there is a deadline put in place to both get the actions realised (done) and checked that they are done.

This company clearly cares about the health, safety and wellbeing of its workforce. It can easily demonstrate to a health and safety inspector that it has taken **reasonably practicable** steps to make sure the machinery is as quiet as possible; that is, it hasn't spent vast sums of money on new machinery or appointed costly acoustic consultants to advise on noise transmission readings. It has clearly taken reasonable steps to make sure that the workplace is the best it can be with the resources available. This is likely to:

✓ Make the workplace safer.

✓ Keep the workforce productive.

✓ Demonstrate to the workforce that the employer cares about their health, safety and long-term wellbeing.

What are the hazards?	Who might be harmed and how?	What are you already doing?	What further action is necessary?	Action by whom?	Action by when?	Done
Noise	Staff and others may suffer temporary or permanent hearing damage from exposure to noise from woodworking machinery.	• Noise enclosures used where practicable, and kept maintained in good condition. • Low-noise tooling used where possible. • Planned maintenance programme for machinery and LEV systems. • Suitable hearing protectors provided for staff and staff trained how to use them. Check and maintain them according to advice given by supplier. • Staff trained in risks of noise exposure and in systems of work to reduce noise exposure (e.g. suitable feed rates for certain job, timber control, etc.).	• Consider if certain machines could be safely mounted on anti-vibration mountings. • Include noise reduction in specification for new vertical spindle moulder, to be purchased next year.	Manager Manager	30/10/20 30/10/20	29/10/20 29/10/20
Vehicles	Staff may suffer injury.	• Forklift maintained and inspected as per lease contract.		Manager	30/10/20	29/10/20

Source: Contains public sector information published by the Health and Safety Executive and licensed under the Open Government Licence.

LO3 Understand how to minimise risks to health and safety

AC3.1 Explain existing health and safety control measures in different situations

Policing health and safety matters

Do you have a friend or family member who is accident prone or needs a little extra help doing day-to-day tasks? Maybe someone you know always seems to have a bandage or plaster-cast on?

The fact is that we are all different and we all need help, even with the simple things in life. Just imagine attempting to carry a hot pizza home from the deli on a china plate – you need help, right? That's why we have packaging, like a box, simple! Well this packaging is just another form of control measure. It prevents you from getting burnt and protects your pizza.

SKILLS // The pizza box is a good example of an everyday and highly efficient control measure. It took considerable resources to make the box a reality. Let's recognise what a control measure is and understand the value it has. A control measure generally has **five simple** categories and the pizza box applies to every one of these. If one of these categories applies to something, then it can be classed as a control measure.

1 **Elimination:** The deli allowed the pizza to stand for a few minutes before placing it in the box, eliminating the chance of a hand being burnt.

2 **Substitution:** The china plate was substituted for a corrugated cardboard box and special cheese selected for its rapid-setting character was used.

3 **Engineering:** The box was designed by a specialist packaging engineer from sustainable materials and the pizzas are a consistent size and shape.

4 **Administration:** The deli manager has ordered a regular and consistent supply of boxes from a reliable manufacturer, with 'Caution hot food' printed on them.

5 **PPE:** The box acts as a device that protects your hands from being burnt by the hot pizza. The deli workers use safety mitts when handling all hot food.

Customers are pleased to be able to safely purchase, transport, share, eat and enjoy their pizzas. Once finished, they can recycle the sustainable packaging and look forward to their next treat. The experience has been a positive and productive one, for both the customer and the deli.

KEY TERM

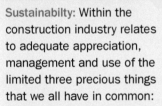

Sustainabilty: Within the construction industry relates to adequate appreciation, management and use of the limited three precious things that we all have in common:

1 Health: For example, the mental/physical health of an individual, commercial health of a business or the wellbeing of both.

2 Time: For example, programme, time constraints and limited or restricted opportunity to access the works.

3 Resources: The availability of materials, labour, plant, minerals, water, education and skills.

(All of the above are non-exhaustive.)

KNOWLEDGE The control measure was put in place by the deli. The construction industry leads the way in putting control measures in place to prevent harm on site. The construction industry has an almost infinite number of different tasks and activities happening at any one time, so site managers and supervisors ensure that the following five control measures are in place for all activities:

| Method statements | Safe systems of work | Work permits | Competent persons | Personal protective equipment (PPE) |

UNDERSTANDING The five control measures are generally applied in the same order of priority, as stated, starting with the drafting of a comprehensive method statement and concluding with PPE. This is because the reliance on PPE is absolutely the last option; planned and preventative measures should always be considered as priority as it is best to avoid, plan, engineer or process your way out of having to rely on PPE.

'Health and Safety PPE Update', https://www.youtube.com/watch?v=B3EDMc_yYWM.

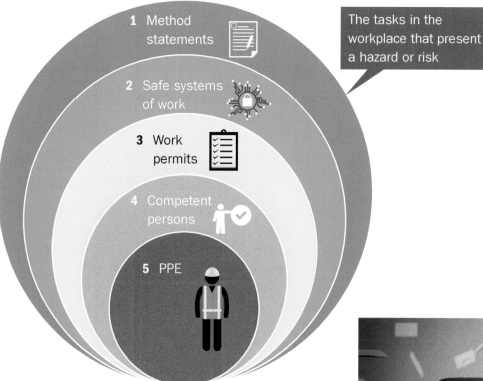

1 Method statements

2 Safe systems of work

3 Work permits

4 Competent persons

5 PPE

The tasks in the workplace that present a hazard or risk

The hierarchy of control measures is intended to minimise exposure of the workforce to health and safety hazards and risks in different situations by consistently applying the five control measures illustrated above. They act like a shield that minimises the reliance on PPE.

The control measure shield

1 Method statements and risk assessments

Written documents that control specific health and safety risks, usually following the preparation of risk assessment that has identified a significant or specialist activity such as demolition, lifting, WAH, using plant, dismantling and complex activities

They should be written by competent persons, who are suitably qualified and experienced in the task that they relate to. The content of the statement will vary depending on the task, but they often have similar elements such as: who, what, where, when, how, stakeholders, duration and a step-by-step description of what is to be done.

2 Safe systems of work

Safe systems of work are formal and recognised procedures that specifically relate to a systematic examination of a task. They provide safe methods to ensure that risks are minimised/ mitigated as far as reasonably practicable. They are often written by professional bodies, institutions, manufacturers of specialist plant or trade associations. The law requires them to be planned, organised, performed, maintained and revised as appropriate. They should be precise and understandable to the workforce.

5 PPE

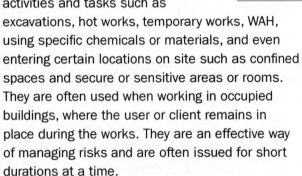

Hard hat, protective boots, high-visibility vest, protective gloves, safety goggles/glasses, dust mast, ear defenders, safety harness, life jacket, welding mask.

Control measures are a proven means to protect the workforce and those that might be affected by construction work or activity. Appropriate control measures should reflect the specific restrictions that may be unique or a requirement of a specific situation.

4 Competent persons

'Competent persons' is the term given to the suitability of workers to perform the role and assume the responsibility or function that they perform on site. They should always be suitably qualified and experienced workers, who often form safe and productive teams of varying sizes. Everybody on site should be 'competent': Architects, Labourers, Engineers, Managers, Bricklayers, Plumbers, Groundworkers, etc. Competency is often recognised by academic vocational and touch-screen aptitude-type certification and testing. These tests are often repeated in short cycles to ensure skill, knowledge and understanding are both relevant and current.

3 Work permits

Work permits are procedural time-constrained documents that are issued by site managers and supervisors to the workforce for the execution of specifically hazardous activities and tasks such as excavations, hot works, temporary works, WAH, using specific chemicals or materials, and even entering certain locations on site such as confined spaces and secure or sensitive areas or rooms. They are often used when working in occupied buildings, where the user or client remains in place during the works. They are an effective way of managing risks and are often issued for short durations at a time.

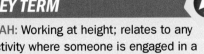

KEY TERM

WAH: Working at height; relates to any activity where someone is engaged in a task that has a risk of falling (from any height) that could result in harm.

Any compliant organisation will have a defined system of company policies and procedures that establish their own rules of conduct and explain the responsibilities of both employees and employers. **These company policies and procedures are put in place to protect the rights of workers as well as the business interests of employers. The shield above forms a strong defence against risk to the business.**

AC3.2 Recommend health and safety control measures in different situations

The construction site is a hub of activity that can **rapidly change both in appearance and in its daily routine**. This is because safe and productive construction environments are the objective of any business that desires safe, timely and cost-effective outcomes. The very nature of construction planning and programming demands that the project often moves rapidly and so changes in form and function. **Safe and productive sites work collaboratively** when the managers, supervisors and operatives effectively apply **suitable control measures** to the relative **situation**. All stakeholders should accept that there are very few 'permanent conditions' on a productive construction site.

REMEMBER

No condition is permanent ... change happens.

SKILLS We need to be able to appreciate how the correct control measures ❶ need to be recommended in the relative situation ❷ to understand how to minimise the risks to health and safety. By having effective situational awareness ❸ we are able understand and conclude that the workplace is safe.

REMEMBER

These are the elements that influence situations. These situations are relevant to individuals and businesses alike.

 Locations

 Changes in work practices

 Equipment

 Scale

KNOWLEDGE The individual and the business both have a responsibility to plan, manage and implement control measures in different situations by maintaining a constant dialogue of interaction and engagement with one another. This is referred to as employer and/or employee engagement, and is once again testament that **'safety is everybody's responsibility'**.

On a construction site it is vital that control measures are considered and implemented through administration procedures by site management that ensures control measures are carried out by the workforce, controlled by supervisors and checked by the site management (see example on page 42).

This cycle of compliance is a minimum requirement for any safe and productive organisation while ensuring safe systems of work are in place. When effective situational awareness is applied to the planning and execution of site activity, then risks to health and safety are consistently minimised.

On the following page the diagram shows how the cycle of compliance is affected by administration procedures, site activity and situational awareness

1 Administration proceedures

2 Site activity

3 Situational awareness

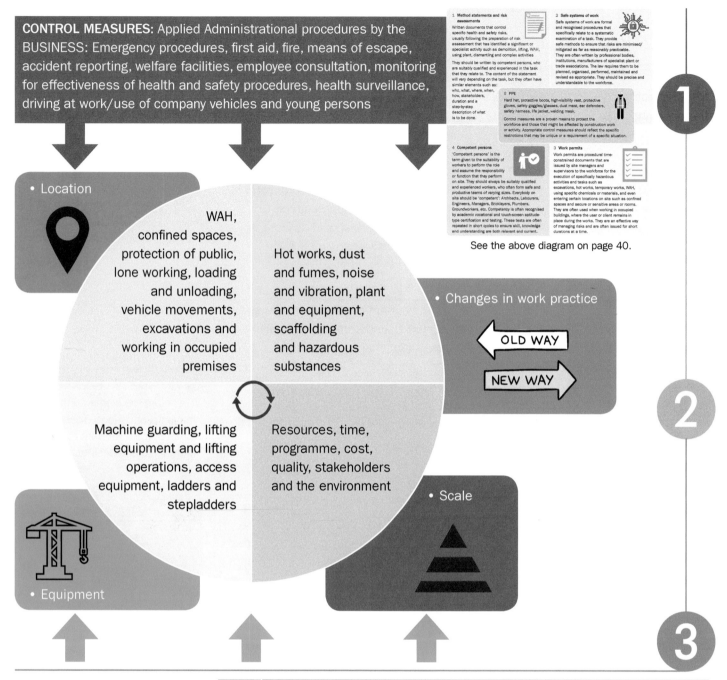

CONTROL MEASURES: Applied Administrational procedures by the **BUSINESS:** Emergency procedures, first aid, fire, means of escape, accident reporting, welfare facilities, employee consultation, monitoring for effectiveness of health and safety procedures, health surveillance, driving at work/use of company vehicles and young persons

1 Method statements and risk assessments
Written documents that control specific health and safety risks, usually following the preparation of risk assessment that has identified a significant or specialist activity such as demolition, lifting, WAH, using plant, dismantling and complex activities

They should be written by competent persons, who are suitably qualified and experienced in the task that they relate to. The content of the statement will vary depending on the task, but they often have similar elements such as: who, what, where, when, how, stakeholders, duration and a step-by-step description of what is to be done.

2 Safe systems of work
Safe systems of work are formal and recognised procedures that specifically relate to a systematic examination of a task. They provide safe methods to ensure that risks are minimised/mitigated as far as reasonably practicable.
They are often written by professional bodies, institutions, manufacturers of specialist plant or trade associations. The law requires them to be planned, organised, performed, maintained and revised as appropriate. They should be precise and understandable to the workforce.

5 PPE
Hard hat, protective boots, high-visibility vest, protective gloves, safety goggles/glasses, dust mast, ear defenders, safety harness, life jacket, welding mask.
Control measures are a proven means to protect the workforce and those that might be affected by construction work or activity. Appropriate control measures should reflect the specific restrictions that may be unique or a requirement of a specific situation.

4 Competent persons
'Competent persons' is the term given to the suitability of workers to perform the role and assume the responsibility or function that they perform on site. They should always be suitably qualified and experienced workers, who often form safe and productive teams of varying sizes. Everybody on site should be 'competent': Architects, Labourers, Engineers, Managers, Bricklayers, Plumbers, Groundworkers, etc. Competency is often recognised by academic vocational and touch-screen aptitude-type certification and testing. These tests are often repeated in short cycles to ensure skill, knowledge and understanding are both relevant and current.

3 Work permits
Work permits are procedural time-constrained documents that are issued by site managers and supervisors to the workforce for the execution of specifically hazardous activities and tasks such as excavations, hot works, temporary works, WAH, using specific chemicals or materials, and even entering certain locations on site such as confined spaces and secure or sensitive areas or rooms. They are often used when working in occupied buildings, where the user or client remains in place during the works. They are an effective way of managing risks and are often issued for short durations at a time.

See the above diagram on page 40.

• Location

WAH, confined spaces, protection of public, lone working, loading and unloading, vehicle movements, excavations and working in occupied premises

Hot works, dust and fumes, noise and vibration, plant and equipment, scaffolding and hazardous substances

• Changes in work practice

OLD WAY

NEW WAY

Machine guarding, lifting equipment and lifting operations, access equipment, ladders and stepladders

Resources, time, programme, cost, quality, stakeholders and the environment

• Scale

• Equipment

1

2

3

SKILLS Look at the task at hand.

By applying the principle of 'SLAM' we can proactively or reactively understand the measures that need to be used to minimise the risk:

 Stop: Engage your mind before your hands.

 Look at your workplace and find the hazards that may impact on you and your colleagues. Report your findings to your supervisor.

 Assess the effects the hazards may have on you, work, procedures, pressures, colleagues and the environment. Do you have the knowledge, training and tools to do the task safely?

 Manage: If you don't feel safe then stop working, tell your supervisor what you think the issues are.

 REMEMBER

Situational awareness is one of many behavioural safety techniques that is useful and easy to use. The SLAM technique promotes a proactive approach that can be applied before, during and after the work. See the next page for an example of a typical day.

Start of employees' day

1 07:30 The employees will unlock their secured vehicle and drive it from their company's secure compound that is protected by CCTV and remote monitoring systems. Employees will often travel to site in pairs to reduce carbon emissions and mitigate 'lone' working. They will make regular check-calls to their employer during the course of the day so that their location is known at all times and they are never left alone and vulnerable.

2 08:00 The employees arrive at their new site and have an induction from the main contractor. They ensure that all their details are kept secure on the main contractor's IT system and that their CSCS and qualification details are handed back to them. They also ensure that the vehicle is parked in a safe and authorised location on site and the keys are kept safe.

3 08:20 After the induction the employees lock their possessions in the secure lockers provided by the main contractor. The locker room is kept locked and access is via a key-coded lock.

4 08:30–10:00 The employees ensure that their tools remain safe at all times and within line of sight to prevent unauthorised borrowing or theft. Their tools will also have their unique employee ID number, postcode or mobile telephone number marked/engraved on them to deter theft and aid recovery.

End of employees' work day

12 It has been a safe and secure shift, everybody has stayed alert, followed procedures, locked their possessions and tools away when not in use and reported unusual occurrences relating to security and safety. They will ensure that their company vehicles remain secure overnight and alarm systems are engaged. They understand these security systems will protect the best interests of their employer and themselves.

Break

5 10:00–10:20 During morning break time the employees ensure that their tools are not left unattended while they rest. Some sub-contractors 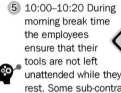 alternate their break times so colleagues can watch tools and materials in the workplace and vice versa. Cutting tools and fuel are locked away and not left out during this time. Everybody on site remains alert.

STAY ALERT

6 10:20–13:00 The employees have been asked to go to the site stores and then the builders merchant. They ensure that their vehicle is locked and that they have the correct order numbers and purchase codes from their employer to ensure that the suppliers understand what materials are to be legally obtained and optimise the time off site.

Result: Criminals are prevented from stealing from site as the collaborative workforce are well trained and highly efficient at loss prevention.

Stop – Engage your mind before your hands. Look and think how security could be compromised. Always stay alert.

Look – At your workplace and find the security hazards that may impact on you and your colleagues. Report your findings to your supervisor. Look for the gaps!

Assess – the effects of the hazards may have on you, work, procedures, pressures, colleagues and the environment. Can theft, vandalism or privacy occur? Secure valuables when they are left unattended.

Manage – If you or your environment doesn't feel secure or if you suspect there could be an issue then stop working and tell your supervisor what you think the issues are.

11 15:00–18:00 The employees will be on site for another week, so the main contractor provides them with a secure store on site to store their tools overnight, this helps minimise manual handling too. The employees secure their tools and materials in the store ensuring it is locked prior to leaving site. When contractors are working in areas where the client is present, they check all tools are accounted for to prevent harm to building users.

Break

10 13:30–15:00 During the afternoon period everybody on site stays alert to potentials of breaches to security. Mobile communications and CCTV help everyone to efficiently communicate with one another; this is particularly important on larger sites where lines of sight can be many miles rather than metres.

STAY ALERT

9 The Site Engineer notices a vehicle driving across the car park very slowly. The Engineer does not recognise the vehicle!

Lunch

8 13:00–13:30 During dinner the employees ensure that their tools are not left unattended while they rest. Some sub-contractors alternate their break times so colleagues can watch tools and materials in the workplace and vice versa. Cutting tools and fuel are locked away and not left out during this time. Everybody on site remains alert.

7 Note: During the course of the day plant operators ensure that the keys for their vehicles are never left in the vehicles. The vehicles are locked when not in use. They will report any suspicious activity or behaviour to their supervisor or the site manger.

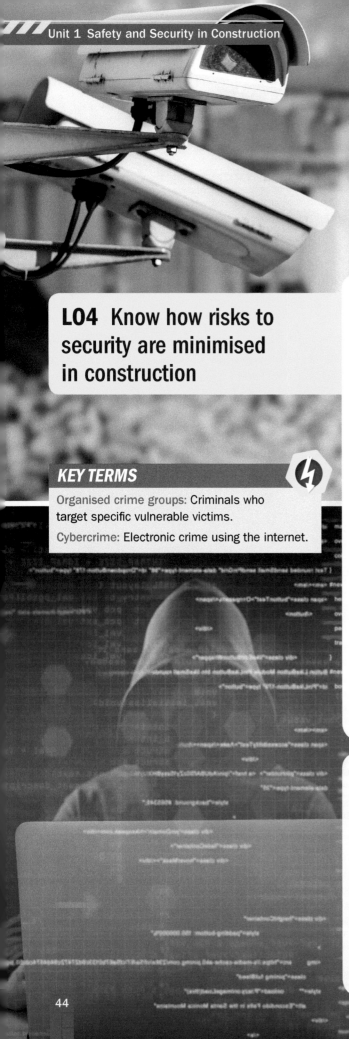

LO4 Know how risks to security are minimised in construction

AC4.1 Identify risks to security in construction in different situations

Construction activity in the UK is worth approximately 8% of everything the country produces. It is no surprise that the industry is targeted by both criminals and small and large organisations, some of which are well resourced and managed.

These criminal groups are known as organised crime groups. Collectively, these criminals target specific vulnerable victims with the intention of stealing expensive mechanical machinery and plant, portable tools, hand tools and even everyday belongings such as PPE and personal possessions from site accommodation, locker rooms and secure containers.

More recently, criminals steal sensitive information such as electronic data, identification, national insurance and tax details, as well as telephone numbers and addresses. This private and often confidential information (data) is sold on or traded for money or more stolen items, with the intention of committing further cybercrime.

Cybercrime is a separate and often invisible crime that is measured separately but is as equally devastating to both the construction industry and the wider world.

SKILLS // When planning, managing or working on site you should be able to identify the risks to the construction industry in different situations. Just like you identify safety risks, you should be equally skilled at identifying the hazards deliberately created by criminals that can be as devastating as a safety-related accident or incident.

FACT

More than 17 million UK citizens were impacted by cybercrime in 2017. (Source: Alex Hern (2018, 23 January) 'Cybercrime: £130bn Stolen From Consumers in 2017, Report Says', the *Guardian*)

APPLICATION / Copy the following table into your notebook and complete it by inserting the current value (in £) of the items by researching the cost of the product when new. Then tick the location where they would be used or stored on site. The equipment can be in more than one place.

Item	Value (£)	Location where used or stored on site			
		Canteen	Workplace	Office	Vehicle
JCB 13-ton tracked excavator					
Cherry picker with more than 15m 'reach'					
24V battery-powered hammer drill					
10 handsaws					
25 full-sized plasterboards					
1 ton of individually bagged cement					
2 personal computers					
Pair of short-wave radios					
1 mobile telephone handset (high specification)					
1 hand portable toolbox full of hand electrical tools					

UNDERSTANDING /

SCENARIO: *Howlyn Construction is a small family owned, multi-skilled company that responds quickly to the needs of its customers by attending to emergency callouts to construction projects where houses are under construction and often behind schedule.*

Howlyn attends a busy construction project with over 100 workers on site to help with the completion of a luxury home to complete some carpentry, minor decorations and fit expensive fixtures and fittings. There are other contractors working in the luxury house. Howlyn has sent a site manager, Pippa Davies, supervisor, Rhys Williams, and six multi-skilled operatives to the site. On arrival, Pippa and Rhys complete a security checklist prior to starting work to ensure the security of their staff and Howlyn's interests.

Working as a team, Pippa and Rhys survey the site and look for the risks and record them on their survey sheet, as shown on the next page. This sheet is then issued to the main contractor for them to take the necessary actions to ensure the security situation is improved.

ID	SECURITY RISK			Description of RISK
	Tools and equipment	Personal belongings	Sensitive information	
1			✓	After the induction, Rhys noticed that the site manager put the induction folder with the personal details of Howlyn staff on the canteen table and left it unattended.
2			✓	Pippa noticed that the site manager did not have a password or other security measure on his laptop and left it unattended in the site office.
3	✓	✓	✓	Rhys observed workers going in and out of the site office taking and returning keys for the properties under construction without signing a register. The workers just open the manager's unlocked desk when they want to and without asking. Some of the houses on site are already occupied and the residents are not at home.
4		✓	✓	The canteen has only ten lockers and there is no facility to secure them.
5		✓	✓	The lockers appear to be former storage boxes that have been fastened together with screws.
6		✓		Pippa inspected the drying room and there is no lock on the door and no coat hooks present; she wondered how workers manage to dry their wet clothes and how their belongings don't get misplaced or mixed up.
7	✓	✓	✓	Rhys observed that the site car park only has ten spaces, so the workers must park 500m away on a disused and derelict site that does not look safe or secure, the manager tells Rhys not to worry so much as he is insured.
8	✓	✓		Pippa could not see any form of secure container on site to store some materials overnight. The main contractor is responsible for storage on this project and states that the site is safe and how they 'haven't had much trouble to date'. There is no security guard on site.
9	✓	✓		The main gate of the site is not locked or secure and vehicles appear to be coming and going as they please without being challenged.
10	✓	✓		Rhys observed that there are tool chests outside most of the houses left to complete, but the locks are combination type and they all have the same code of '1234' and workers from different companies share access to the tool chests.
11	✓			Pippa records the fact that mini-digger and dumper operators have gone on break and left the keys for these machines in the ignition and parked by the entrance gate. They are unattended for 30 minutes. The dumper driver leaves his identity card on display and hanging from the keyring. There are several hand tools also on display in the storage of the tray of the excavator.
12	✓			There are several hand tools left unattended adjacent to the trench where the groundworkers are working.
13	✓			Pippa can see that the tower crane has no lock on its barrier gate at ground level. There is a group of young adults pointing at it through a damaged section of open mesh-type fencing.

KNOWLEDGE Pippa and Rhys issue the survey sheet to the site manager and report their findings to Howlyn's head office. They also advise Howlyn not to work on the site as it is not secure and considered to be a considerable risk. They are commended for their excellent situational awareness.

The main contractor must compensate Howlyn for non-productive work as their site is not secure because there are so many security failures. The project falls further behind schedule. It is likely that incidents of theft will increase.

UNDERSTANDING Construction-related data and information are often highly confidential, so this too should be kept in a secure environment. It includes documents and data such as:

- contracts
- letters and emails
- specifications and drawings
- access to data storage websites and common data environments (BIM)
- stakeholders' information (clients', designers', sub-contractors', operatives', suppliers' and customers' intellectual property such as concepts, designs and literature).

This is because many clients and other stakeholders demand that their business and their business interests remain tightly controlled and managed to prevent theft, vandalism, crime, cyberattacks, terrorism and potential threats from organised crime groups. For example:

1. The Ministry of Defence (MoD) will want to protect its design for a new military base from getting into the possession of enemies of the state or other foreign powers.

2. A bank under construction will not want the design of its new vault and security system to be common knowledge.

3. A resident in the community who has adaptations made to their house for special medical needs may wish to keep this private.

4. A contractor competitively tendering to win a multi-million-pound contract will not want their competitors stealing their ideas and concepts.

APPLICATION First, find out what GDPR relates to and summarise it in your notebook.

Then, download from the dedicated website (see QR code on the right) or copy the risk table on the next page into your notebook and identify the potential areas of what could be lost in terms of money, time (productivity), data and physical things such as possessions. Tick all the relevant boxes.

Downloadable

Risk to security	£	Production	Property	Data

Contracts

Letters and emails

Specifications and drawings

Access to data storage websites and common data environments (BIM)

Stakeholders' information (clients, designers, sub-contractors, operatives, suppliers and customers)

Intellectual property such as concepts, designs and literature

AC4.2 Describe measures used in construction to minimise risk to security

Just like the technique used for applying control measures to prevent accidents and incidents, the same theory can be adopted to minimise risks to security of the construction process. This means that security, just like safety, is often everybody's responsibility. This is because a breach in security could easily result in an accident or a theft that could result in harm or delay a construction process or project.

These measures are used collaboratively by both employees and employers to protect the built environment from theft, vandalism, trespass and the odd nosey member of the public.

SKILLS // By applying the same principle of SLAM as in AC3.2, we can proactively or reactively understand the measures that need to be used to minimise the risk to security:

LINK

For more on SLAM see page 42.

 Stop: Engage your mind before your hands. Look and think how security could be compromised. Always stay alert.

 Look: At your workplace and find the security hazards that may impact on you and your colleagues. Report your findings to your supervisor. Look for the gaps!

 Assess: The effects the hazards may have on you, work, procedures, pressures, colleagues and the environment. Do you have the knowledge, training and tools to do the task? Can theft, vandalism or privacy occur? Secure valuables when they are left unattended.

 Manage: If you or your environment doesn't feel secure or if you suspect there could be an issue then stop working. Tell your supervisor what you think the issues are.

Now you can describe the measures used by employees and employers to minimise risk to security.

Developing Construction Projects

LO1 Be able to interpret technical information

AC1.1 Interpret technical sources of information

The interpretation of technical information is a vital skill that enables stakeholders in the construction industry to communicate, understand and comprehend simple and complex information and concepts. These construction-related concepts and designs can be made up of freehand sketches, simple ruled lines, watercolour washes, formalised and regulated designs, schemes and plans. More recently, detailed and data-rich digital models, renderings and graphics, generated by powerful software and computing, have revolutionised how the industry communicates, shares, recovers and uses design information.

That said, regardless of how it is generated and shared, the information must be capable of being understood, often by a broad range of stakeholders with an equally broad range of capabilities. Just like the extensive use of standardised pictograms within construction signs, the construction industry uses a broad standardised series of graphical symbols, signs, marks, words and narratives to help define the concept that is being illustrated.

This is because different occupations often need to see different aspects of the concept, for example:
- A developer may only need to view a potential site they want to purchase in context of the adjacent areas, roads, communication links, floodplains, etc.
- A site manager may need detailed layouts of all aspects of a floorplan so they can understand the sequencing of various trades and implications of site logistics.
- An electrician may only need a simple wiring diagram to complete a control panel wiring exercise for a highly complex electrical system.

So, the possible content and uses of a drawing are infinite.

REMEMBER

Any drawing or document (or information) that does not contain a recognisable symbol of reference should contain a legend or key to allow the reader to verify what the symbol or narrative means.

SKILLS / Technical information

In construction terms, though, you not only need to communicate with each other by using understandable drawings and concepts such as 3D models but also interpret specifications, building regulations (including other standards) and design briefs. Now that you have developed the health and safety skills (see Unit 1) to potentially become safely productive, we can use the interpretation skills from this unit to select and order materials, tools and equipment to effectively and economically realise and complete tasks.

KNOWLEDGE / There are numerous sets of standards related to the presentation of this data (a detailed knowledge of this **is not** part of this qualification). However, there are three primary elements that help stakeholders to interpret information within the built environment:

- symbols
- conventions
- terminology, e.g. 'part of the works'.

REMEMBER

It is vital to know that these three elements frequently vary in format. It is often the case that different organisations and companies have their own approach to some or all of the elements.

Symbols

- Below is a sample selection of symbols that are frequently used in the construction industry as a common form of reference.
- Remember that these often vary slightly, depending on an organisation's own procedures.
- The meaning of these always will be illustrated as part of best practice to allow the reader to fully understand their meaning.

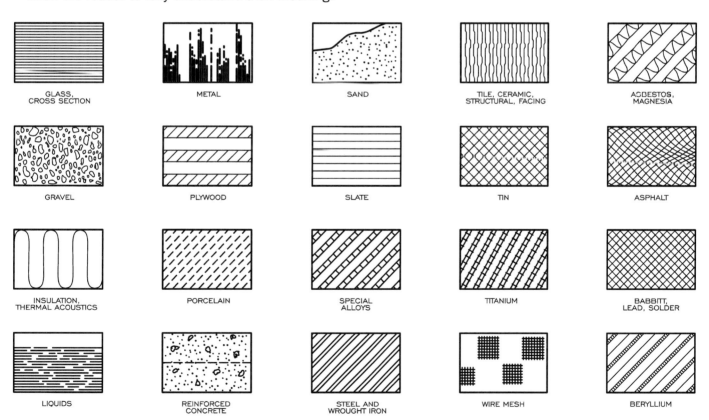

Architectural and graphic symbols/symbols for materials

KEY TERMS

Aesthetic: In the construction industry, the term 'aesthetic' is used to describe the visually pleasing standards of a building, parts of building or overall impression of a particular detail of the building that has been formed. This pleasurable and descriptive term is often used when contrasting details such as: 'The black and grey natural Welsh slated conical roof turret contrasted with the green leaves of the trees behind the building. It was aesthetically pleasing to see the building in context of the adjacent valley.'

Impression: In the construction industry, clients, through their design team, often refer to the word impression to relay a concept or feeling of a building or space within the built environment. For example, 'The castellated stonework gave the impression of a castle wall and so made the building's envelope feel robust and more secure.'

Conventions

- A series of accepted, traditional graphical aspects and impacts of the design process that have developed over, sometimes, hundreds of years and relate to how design data is summarised in:
  - plans
  - section
  - elevations
  - isometrics
  - vanishing point drawings
  - freehand sketches.
- It also refers to summary narratives in table form relating to when and how the drawing was drafted and revised, and by who. These often vary slightly, depending on an organisation's own procedures but generally have the same key data.

Terminology

- Like conventions, construction terminology has evolved over many years and contains a mix of prescriptive legal-sounding words such as must do and must not narratives, together with softer but definitive aspirational descriptions such as aesthetic and impression.
- This is because many clients want their building to be individual or have an expectation of how it will be constructed or what it will look like.
- The terminology often reflects the same use of language as the main contract, hence its formal-sounding content.

What documents do we need to be able to interpret?

Specifications

- Prescriptive documents that are generally commissioned by the employer to summarise the anticipated materials and associated workmanship standards that the employer desires.

- Specifications often form part of the contract and are vital documents in the construction process.

- They are intended to set the standard for the construction phase of a project.

- They rarely relate to quantities or cost. Some specifications are known as performance specifications and relate to a construction product's or building's performance or characteristic as opposed to an employer's strict perception of what the materials or method should be.

- Specifications often reference specialist and traditional specific products such as a roof tile, brick or paint covering. Specifications are often written in a sequence that reflects the primary elements of a building such as substructure, superstructure, envelope, external works, etc. and consider the vital interfaces.

KEY TERM

Interfaces: Where and how things relate to one and other.

Drawings

- Drawings are produced by the design team and specialist contractors to illustrate concepts, designs and what could be possible to achieve.
- They form a vital and pivotal suite of documents that can be used in conjunction with all other contractual and relevant documents.
- Drawings are often issued following a strict timeline of quality and scope known as the RIBA plan of works. This plan of works identifies key periods leading up to, during and after the works and is the product of decades of good and bad experiences together acting as a record of lessons learnt.
- In most cases all other designers such as Structural and Civil Engineers, mechanical and electrical specialists and specialist contractors follow this plan in a collaborative effort to ensure all stakeholders deliver the overall design in a timely fashion (time to make changes if necessary). In doing so the project is more likely to be safer and more productive, as everyone has a chance to understand and comment on the completed design.

Building Regulations

- The Building Regulations form the set of institutional standards that buildings are constructed to in the UK.
- These standards are often updated to take into consideration the rapidly changing and developing standards and demands of society to reduce the consumption of fuel, power and waste.
- The standards are enforced by the local authorities which are supported by central government in ensuring they are imposed, and that buildings comply to some of the highest current standards of workmanship and materials.
- The Building Regulations have existed since the mid-1960s and continue to develop, helping the UK to produce safer, more efficient and more sustainable buildings to cope with a rapidly demanding and growing population.
- These regulations often apply to the refurbishment of existing buildings to ensure all building stock is consistently produced and used efficiently.

Design brief

- The design brief is a document often produced in the format of a report.
- This brief is produced by the client or lead designer to summarise key aspects and aspirations of the project.
- The brief often contains the following:
  1 Reference to the client
  2 Site information
  3 Spatial requirements
  4 Technical aspects
  5 Material and component requirements
  6 Project requirements.

These are all discussed on the following page.

1 Reference to the client

What they do and what the building is to be used for, to provide a greater understanding of the end purpose of the project.

2 Site information

Including surveys, limitations and restrictions that may help or hinder the concept or feasibility of the project. Legal and planning matters are also featured to ensure statutory compliance.

3 Spatial requirements

Including dimensioned plans and elevations, including any special circulation or access requirements that could be compromised by other design features or when the building is furnished or fitted out. Key aspects such as temperature, zoning and phasing considerations are also considered here.

4 Technical aspects

This is often the most comprehensive section of the report, as all the technical data that is the relevant to the project is summarised, including:

- Structural strategy (type of new and existing structure, gridlines, floor-to-ceiling heights and special loadings such as racking or vehicles).
- Anticipated lifespan of the building or refurbishment.
- Servicing requirements (scope and frequency of specialist servicing and maintenance).
- User controls and levels of comfort (cooling, heating and ventilation).
- Acoustic requirements (segregation and soundproofing).
- Specialist equipment and plant requirements (furniture, fittings and fixtures, including IT).
- Flexibility requirements and future proofing.
- Waste, pollution prevention and security requirements.
- Energy use (including special considerations for reduction of consumption of fuel and power).
- Other (any other special client requirements).

KEY TERM

Future proofing: How the building can adapt and cope with change in the future.

5 Material and component requirements

Materials and products that are of significance are listed or referred to. Similarly, products that are not allowed for special reasons can be summarised here also. This is because some clients or organisations have a reason for specifying them.

6 Project requirements

This section has key project delivery data such as named contractors or designers, budgets and cost summaries, planning issues and conditions, timeframes, and programmes and tendering requirements.

KNOWLEDGE // BIM

Building information modelling (BIM) has rapidly become a digital one-stop design, engineer, construct and maintain aspirational concept. This approach fully embraces and includes all the sources of information in one digital model. The aspiration to have various levels of BIM and its use throughout the industry is making slow but steady progression as stakeholders learn to fully embrace its introduction and use. This is partly due to the complex software interface and platform challenges that different products and users experience. It is also partly due to the reluctance of the industry to invest in costly hardware and software licenses that currently do not fully deliver what is expected of them. The UK government has set targets for its introduction and regularly reviews progress made.

AC1.2 Plan sequence of work to meet requirements of sources of information

By failing to plan, you are planning to fail! By failing to plan, a sequence of work is potentially hazardous, unproductive and ultimately unsustainable. When you plan sequences of work in the construction industry, however simple or complex, you are **likely to have a safer and more productive outcome**. This makes the chances of achieving a greater profit more likely too. Just imagine all those trades, occupations and construction projects working without a planned sequence of work. When the industry plans effectively and collaboratively, the country's economy operates more efficiently and is better placed to cope with fluctuations in labour shortages, material procurement delays and increased demand for services.

SKILLS It is vital that you possess the skill of being able to understand the full suite of requirements that have to be achieved from your sources of information, such as **specifications**, **drawings**, **design briefs**, **building regulations** and **stakeholders'** resources in a logical series of planned events; ideally, you can prepare, install, test, inspect and sign off the work as you complete it. You can then apply for payment for your work and continue to be productive. Now you will learn how to effectively **plan sequences of work to meet the requirements of the sources of information**.

KNOWLEDGE

1 Remembering that **health and safety** is your number one production priority, you should consider **HOW** your actions will affect your plan of work.

2 Considering **WHEN** and **HOW LONG** producing the work will take, you must give adequate time to do so and factor in float (spare time), just in case, to conclude a feasible **timescale**.

3 Understanding in what **ORDER** the activities follow each other is vital to conclude the correct **sequence**.

This knowledge now prepares you with an invaluable mechanism to understand how, when, how long and in what order information is processed in a timely and productive manner, in readiness to start the work.

UNDERSTANDING A perfect example of this process in action is provided by one of the oldest crafts and trades in the industry. The design and construction of brickwork masonry, despite its simple appearance, is often fraught with complexities due to its technical, aesthetic and workmanship detailing and non-conformities. When brickwork is designed, constructed and inspected properly, it can last hundreds of years. When the opposite applies, then it can lead to hazardous, imperfect, ugly and abortive construction work that has a short lifecycle, often ending in demolition and costly reworking. The process shown on the next page illustrates how the sources of information can be translated into methods and safe systems of work, and ultimately incorporated into a construction programme.

> **KEY TERM**
>
> Float: Spare/extra time that may be needed.

APPLICATION / The brickwork plan of work will be constructed using the following process.

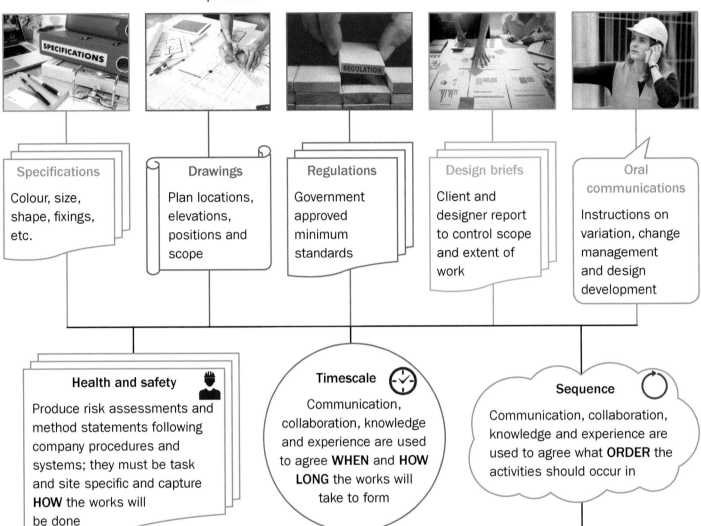

Specifications

Colour, size, shape, fixings, etc.

Drawings

Plan locations, elevations, positions and scope

Regulations

Government approved minimum standards

Design briefs

Client and designer report to control scope and extent of work

Oral communications

Instructions on variation, change management and design development

Health and safety

Produce risk assessments and method statements following company procedures and systems; they must be task and site specific and capture **HOW** the works will be done

Timescale

Communication, collaboration, knowledge and experience are used to agree **WHEN** and **HOW LONG** the works will take to form

Sequence

Communication, collaboration, knowledge and experience are used to agree what **ORDER** the activities should occur in

A construction programme of work is produced and agreed:

Activity	Week 1	Week 2	Week 3	Week 4	Week 5	Week 6	Week 7	Week 8	Week 9
Order materials	◆—◆								
Erect and check scaffolding	◆——◆								
Safely load scaffold		◆—◆							
Form brickwork		◆——————————◆							
Clean, check and snag						◆———◆			
Sign-off brickwork							◆—◆		
Clean site								◆———◆	

Plan health and safety aspects prior to commencing the work

Safety is everybody's responsibility, so how you plan your approach to work is vital. Regulations, laws, certification, procedures and rules all change and develop over time, so, until such time as you are suitably qualified to plan and manage work, employees should always follow the direction of their employer.

The quality and strength of direction will vary between employers, so if you ever feel or know that your direct supervision is not what it should be, then stop work and speak to your employer, supervisor or manager.

The sign below is intended to be easily read and understood and acts as a behavioural change tool reminding us that planning aspects of health and safety need not be overly complicated.

Are you WORK READY for a day at work?

isg

Positive attitude

Hydrated and fed

Understand the task at hand

Adequate and correct PPE

Understand your specific method statement

Understand your site-specific risk assessment

You are constantly aware of how your actions and work may impact on others (on and local to the site)

Understand and 'buy into' your ISG site-specific rules

You know the location of your on-site supervisor

You possess a **CSCS** card and understand your induction

We invest great resources into managing and improving our construction sites, we need your buy in to ensure that a safe and sustainable work ethic exists on this site. Don't just look the part; know your part and understand that we will look below the waterline when assessing your suitability to work with us.

Howard Davies

Following this approach can help define and plan your own performance when working on your own and with others.

APPLICATION Produce your own behavioural change poster and keep it in a safe place. Develop it and check it over time to see if you are abiding by this code of conduct. If not, record your improvements in a diary to keep a record of your own **CPD**.

REMEMBER

When you fully prepare yourself to work effectively, and adequately plan the sequence of work, you become safer and more productive.

KEY TERMS

CSCS: Construction Skills Certification Scheme; an accredited body that certifies construction qualifications.

CPD: Continuing professional development; a diverse range of courses and self-delivered, lifelong learning that professionals often do to maintain a current level of understanding and knowledge of their relative role or job. This approach often helps them attain more formal qualifications and also develop new skills.

LO2 Know preparation requirements for construction tasks

Many luxury brand companies base their corporate ethos on safety, quality and sustainability.

AC2.1 Identify resources required to complete construction tasks

There are few, if any, construction-related tasks that do not require preparation. If preparation is forgotten or ignored, the simplest and most straightforward of tasks can soon become unsafe, costly and unproductive, often leading to a reduced proft margin.

The activity or operation is likely to become unsustainable, leading to conflict with the client and other trades, as well as poor quality. **Good reputations can soon become forgotten and are hard to regain during this period. In some cases, litigation can follow, which compounds the issues highlighted along with unwanted negative press and media coverage.**

When you prepare for a construction task sufficiently, then that task is likely to be safer, more productive and more enjoyable as a result. The more productive you become and the more you enjoy producing good quality work, the **happier your customers become – and your reputation reflects the experiences of your customers**.

This ethos applies to all businesses regardless of size, turnover and location.

'Health and Safety PPE Update', https://www.youtube.com/watch?v=B3EDMc_yYWM.

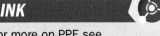
LINK

For more on PPE see pages 15 and 39.

SKILLS By developing the skills to be able to **identify the resources required to complete construction tasks you empower yourself** and others to positively deliver the goods to your customers. When you collectively improve on these skills, wider society and the image of the construction industry improves.

KNOWLEDGE Personal protective equipment

The selection of personal protective equipment (PPE), equipment and tools can be complex and diverse but is greatly simplified and the process is made more efficient by the application of a management process. This process involves the use of a **check sheet** that is prepopulated with options that can be chosen (ticked ✓) or not selected (left blank or ✗). There is also an option to place task-specific notes.

UNDERSTANDING Check sheets should be drafted, read and understood in conjunction with task-specific **method statements** (if applicable) and **definitely with a risk assessment**. This suite of documents will vary in format between organisations, content and detail. All documents should be approved by a supervisor and/or manger prior to starting work.

Although such check sheets are not exhaustive and will need to be revised to reflect any change, they form a robust (i.e. strong) control measure in the pursuit of high standards of **safety, quality and best practice**.

APPLICATION / The example of a check sheet on the next page, which you can download from the website, exemplifies a good example of this process, which can be applied to most tasks or activities within Unit 2.

The check sheet has seven sections which cover all the aspects learners will encounter when preparing for the requirements of construction tasks.

REMEMBER

Documents should also be shared and understood by wider stakeholders who will or may be affected by the work that is proposed before it commences, to ensure opportunity for dialogue and objections can be discussed or raised.

REMEMBER

Download several copies of the check sheet template shown on the next page from https://www.illuminatepublishing.com/built_environment and use them when you plan your next tasks.

The Illuminate Constructing the Built Environment dedicated website, http://www.illuminatepublishing.com/built_environment.

Title of check sheet: Project: Date:	Read this sheet with:	1 Method statements 2 Risk assessments 3 COSHH sheets 4 Other documents

Section 1. Personnel protective equipment (PPE)

Hard hat	Safety boots	Gloves	Safety eyewear	Over-garments	Ear protection	Dust mask	High-visibility clothing	Any other PPE requirements

Section 2. Access equipment (complete with leading-edge protections and toe boards) and temporary protection

Scaffolding	Podiums	Access towers	Step ladders	Ladders (permit required)	Dust sheets	Vacuum cleaner	Foam corner protection	Any other types of protection required

Section 3. Security, existing services and utilities

Keys or codes required	Working hours	Access hours	Welfare and task lighting	Emergency contact details	Existing services locations	Location of water	Location of 110V power	Any other measures needed

Section 4. Tools requirements

4a. Setting out

Pencils and markers	Tape measure/ rulers	Plumb bob	Spirit level	Chalk line	Scribe	String line	Notebook	Other

4b. Hand tools

Screwdrivers and hex keys	Handsaws and blades	Claw hammers and mallets	Spanners, sockets and wrenches	Knives and pliers	Wood and bolster chisels	Staple guns, nail guns and riveters	Files, planes and rasps	Clamps

Bolts cutters and snips	Torches and flashlights	Trowels	Floats	Hawks	Tile cutter ceramic	Brushes, pots and cleaners	Other hand tools	

Section 5. Documentation

Programme	Sequence of work	Specifications	Approved drawings	Building regulations applicable?	Design brief required?	Inspection required by third parties?	Other? (Is a datum or setting out point agreed or oral communications?)

Section 6. Materials, adhesives, sealants and fixings

Textiles	Wood	Brick	Mortar	Plaster	Decoration		Tiling

Electrical	Plumbing	Heritage skills	Other types of materials or fixings required			

Notes on quantities:

Section 7. Storage

Lay date	Covered	Secure	Segregated	Watertight and frost protection (thermometer required?)		Combustible?	Staked or single?

KNOWLEDGE // Materials

The construction industry for some time has relied upon a group of materials derived from natural resources. These mineral- and organic-based materials have in some cases been used for thousands of years to protect humans, livestock and possessions from the elements, predators and sometimes from each other. Without doubt, they have helped humans to develop into what they are today and allowed some nations to thrive. It is not the use of these materials alone that has contributed to advancement. People have learnt to maximise and improve the properties of these materials by advanced manufacturing processes to ensure the best possible value is provided by these materials and products. Modern methods of construction and design now factor-in the recycling of these same materials into new products and even into the building's lifespan, ensuring that selected materials can be re-used when the building is no longer required.

SKILLS // To make the most of these materials you need to acquire the skill of selecting the most suitable and appropriate materials for the task. You have already learnt the value of the specifications, drawings, building regulations and design briefs. Now you have a chance to apply these skills and create a product. This product could be a building component, element or even a complete building or engineering product ready to use or take to the market. So, let's get started!

KNOWLEDGE // You need to know the characteristics (features), qualities (why it's distinctive), sustainability features (the capacity to endure) and limitations (restrictions) of these materials in order to optimise (make the most of) your skills when working with them individually or when formed together. It is an established fact that most building defects occur at junctions of different materials due to lack of comprehension of the properties of materials, caused by poor detailing and construction techniques.

UNDERSTANDING // Selection of resources is limited to your knowledge of their individual characteristics, so it is essential to be able to identify the most appropriate materials for the relative task. These characteristics are represented in the following materials model:

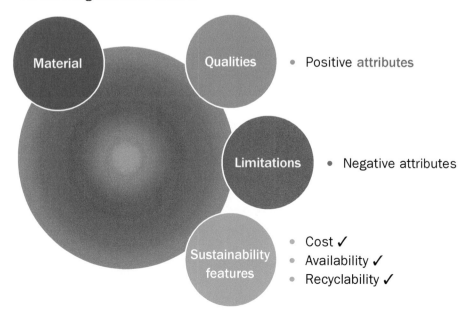

- Positive **attributes**
- Negative attributes
- Cost ✓
- Availability ✓
- Recyclability ✓

KEY TERM

Attribute: Quality or feature.

LINK

For more on working with materials see pages 92–105.

APPLICATION / Let's look at the characteristics of all the materials that are encountered as part of the course and learn how to apply your skills, knowledge and understanding when selecting them.

Textiles

Qualities
- Attractive
- Flexible
- Wide range of uses
- Cost effective

- Help prevent heat loss and draughts
- Good acoustic uses

Limitations
- Low structural uses
- Quickly become dated designs
- Attract dust

Sustainability features
- Cost ✓
- Availability ✓
- Recyclability ✓

Wood

Qualities
- Lightweight
- Moderate compressive and tensile strength
- Multiple uses
- High degrees of tolerance

Limitations
- Known to rot and swell
- Flammable if not protected
- Requires chemical treatments and kiln drying

Sustainability features
- Cost ✓
- Availability ✓
- Recyclability ✓

Bricks

Qualities
- Withstand many years of severe weather
- Modular in nature
- High degree of tolerance
- Strong in compression

Limitations
- Require skilled trades to lay them
- Low tensile strength
- Heavy
- Sustainability features

Sustainability features
- Cost ✓
- Availability ✓
- Recyclability ✓

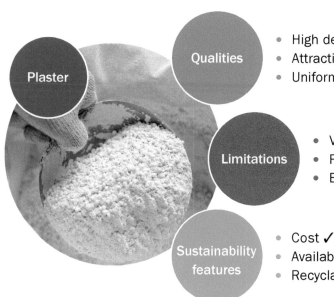

Plaster

Qualities
- High degrees of tolerance
- Attractive
- Uniform finish
- Traditional related trade
- Rapid construction and setting
- Can be decorated

Limitations
- Very poor water and damp resistance
- Requires skilled trades
- Brittle and weak in tension and compression

Sustainability features
- Cost ✓
- Availability ✓
- Recyclability (only by specialists) ✓

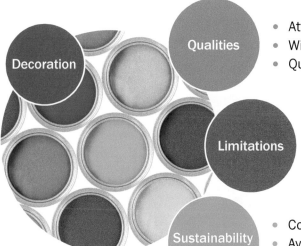

Decoration

Qualities
- Attractive and hard wearing
- Wide range of colours and shades
- Quickly applied

Limitations
- High VOC content when not water based
- Requires several coats
- Requires regular reapplication

Sustainability features
- Cost ✓
- Availability ✓
- Recyclability ✓

Tiles

Qualities
- Long life
- Extensive range of type, colours and finish
- Hardwearing and hygienic
- Semi-skilled trade
- Rapid fixing
- Uniform in shape

Limitations
- Low tensile and compressive strengths, unless specifically manufactured
- Brittle
- Cold to touch

Sustainability features
- Cost ✓
- Availability ✓
- Recyclability ✓

KEY TERM

VOC: Volatile organic compound. Substances, such as formaldehyde, that used to be in certain man-made materials, e.g. MDF or some paints, that are dangerous to health and are now restricted in use.

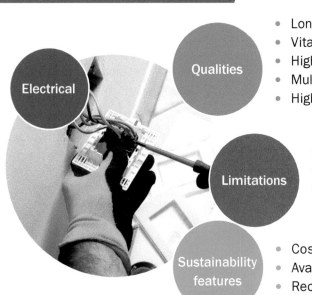

Electrical

Qualities
- Long lifecycle
- Vital to modern society
- Highly regulated
- Multiple uses
- High standards of testing and certification

Limitations
- Requires skilled trade
- Limited resistance to the elements, particularly water
- High risk to installers

Sustainability features
- Cost ✓
- Availability ✓
- Recyclability ✓

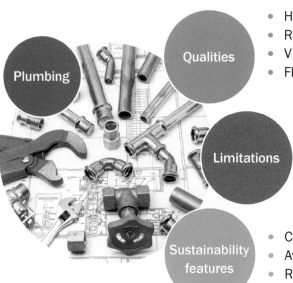

Plumbing

Qualities
- Highly regulated
- Rapidly formed
- Vital to maintain healthy society
- Flexible and modular materials

Limitations
- Requires highly skilled and certified trades
- Requires regular maintenance
- Some long procurement times for special components

Sustainability features
- Cost ✓
- Availability ✓
- Recyclability ✓

Heritage

Qualities
- Traditional appearance
- Natural and locally available materials
- Rapidly constructed

Limitations
- Heavy
- Sometimes difficult-to-reach locations

Sustainability features
- Cost ✓
- Availability ✓
- Recyclability ✓

APPLICATION / Identify the type of materials in the eight images below and link them with the list of nine types of materials. Take note of how the department store and food chain have used the look and feel of the materials as part of their brand identities.

A Textiles
B Wood
C Bricks
D Plaster
E Decoration
F Tiles
G Electrical
H Plumbing
I Heritage

REMEMBER

The materials check sheet on page 62 can help with this process and it should be highlighted that this list is not definitive, and different projects have different situations and challenges.

KNOWLEDGE / The removal and safe disposal of any of these materials requires careful consideration and so should be done with adequate control measures, supervision, permits, risk assessments and methods statements. The location of live services should be known prior to commencing work and the termination of these should also be considered prior to commencing work.

Many contractors now have specific waste (sometimes referred to as 'waste stream') targets that they strive to achieve and improve upon. Many contractors are exceeding 85% of all site waste being recycled. Many sites now have separate waste streams and segregated waste points to make this process more efficient. The waste transfer centres then take this segregated waste and process it once again to maximise the recycled content. At this point waste is even referred to as product, as it is seen as just that. Like most other products it therefore has a value (£).

AC2.2 Calculate materials required to complete construction tasks

When you want to effectively prepare materials, you should have the skills to be able to **calculate the amounts** required to successfully achieve the task. When materials are procured, regardless of type, they will be available in regulated and definitive **weights and/or measures** to ensure that they are consistently manufactured and distributed. You also need to be aware of the limitations of the materials in terms of their working time (wet materials require time to dry) and the ratios in which they need to be blended with other materials/elements.

KEY TERM

Quantification: The end product of a process to conclude the gross (before wages, tax and other expenditure) and net (after paying tax) sum of a particular task or a group of tasks.

LINK

For more on tolerances see pages 82 and 153.

SKILLS / ## Time and cost

The **time** and **cost** of the total process needs to be **accurately calculated to ensure sustainable levels of production can be planned and achieved.**

Their **final weight, size, density and appearance** are often part of the standards to which the materials are processed and/or manufactured to. These same standards are consistently checked by manufacturers, designers, builders and those responsible for checking quality in the industry such as construction managers, Engineers, Architects and all specialist contractors. These checking procedures are governed by **strict tolerances** that are often published as institutional standards and/or approved codes of practice. These are the same standards that are often referred to within specifications, drawings, building regulations and design briefs.

This process of quantification is also an appreciation of the **time** it takes to successfully procure, distribute, prepare and use the materials. So, the process is dependent on **efficient programming** to ensure the series of events from the point or ordering them to the point of arrival on site, to the time when they form part of the completed works, is accurately appreciated.

There are six elemental calculations to help you comprehend the full resources that will be required to complete a construction task:

1 Area
2 Volume
3 Perimeter
4 Time
5 Ratio
6 Costs

When you know how to successfully use these calculations, you are able to fully understand the resources that are required to execute a task. Earlier in this book it was suggested that the only three precious things everyone has in common are: health, time and **resources**. Consider this again.

Before progressing further, please promise yourself that you will not skip this maths-related section, as you will need these skills for the rest of your career.

LINK

For more on health, time and resources see page 31.

The following section has some templates, which you can download from the dedicated website and print onto graph paper. There are two types of graph paper:

Type 1: Standard grid-type graph paper for calculating **area, perimeter, time, ratio and costs** (**costing**). The use of electronic spreadsheets is also ideal for costing.

Type 2: Isometric paper for calculating **volume**.

The Illuminate Constructing the Built Environment dedicated website, https://www.illuminatepublishing.com/built_environment.

KNOWLEDGE To be capable of accurately calculating area and volume we must apply formula. These formulas have existed for many years and remain the basis for Architects, Engineers and builders to accurately understand, plan and communicate designs and concepts in a safe and productive series of numerical values. The formulas are shown on the following page and can be downloaded from the dedicated website and printed out as a reminder.

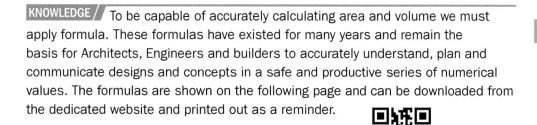

The Illuminate Constructing the Built Environment dedicated website, https://www. illuminatepublishing.com/built_environment.

REMEMBER

In Europe we communicate this data in metres (m) and millimetres (mm), sometimes referred to as 'mil' (slang).

Area formulas

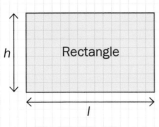

Rectangle

h | l

Area = $l \times h$

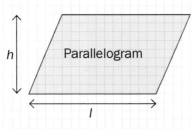

Parallelogram

h | l

Area = $l \times h$

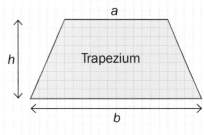

Trapezium

a | h | b

Area = ½ of $h \times (a + b)$

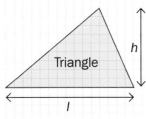

Triangle

h | l

Area = ½ of $l \times h$

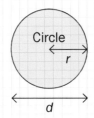

Circle

r | d

Area = πr^2

Key
$\pi = 3.14$

Volume formulas

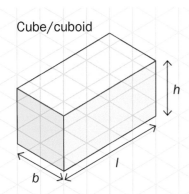

Cube/cuboid

h | l | b

Volume = $l \times b \times h$

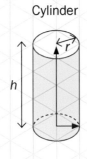

Cylinder

r | h

Volume = $\pi r^2 \times h$

Cone

h | r

Volume = $\frac{1}{3}\pi r^2 \times h$

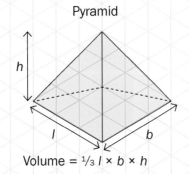

Pyramid

h | l | b

Volume = $\frac{1}{3} l \times b \times h$

Key
$\pi = 3.14$

The effective application of these formulas during the design, procurement and construction processes will influence greater accuracy, quality, production, waste reduction and profits. These formulas are used millions of times a day throughout the construction (and other) industry, regardless of roles and responsibilities.

UNDERSTANDING Let's work through the application of these formulas to gain an understanding of how they surround you in form (shape) and function (as the bay window below). Consider how they would have been designed and realised during the design and construction phase of their lifecycle.

1 Volume

EXAMPLE SCENARIO 1: The bay window shown below right is to be replaced with a new insulated bay window. The Architect needs to know the volume of the bay window to complete a heat-loss calculation. Find the volume of the bay window.

Step 1. Sketch the outline of the bay window on isometric graph paper (it doesn't have to be to scale).

Step 2. Identify the length, base and height.

Step 3. Identify the values (by measuring).

Step 4. Apply your volume formula: $l \times b \times h =$ **Volume.**

Step 5. Conclude your answer.

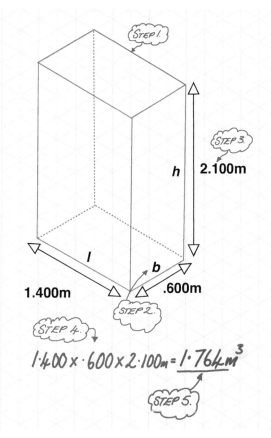

STEP 1.

STEP 3.

h | 2.100m

l

b

1.400m

.600m

STEP 2.

STEP 4.

$$1.400 \times .600 \times 2.100m = \underline{1.764m^3}$$

STEP 5.

SCENARIO 2: This former public house is to be converted into residential units. The client wants to know the overall volume of the building to help with a feasibility plan. Find the volume of the building.

Note: these sketches are not drawn to scale.

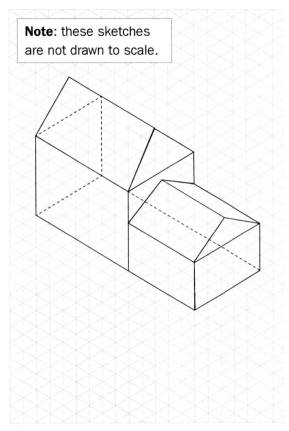

Step 1. Sketch the building on isometric paper.

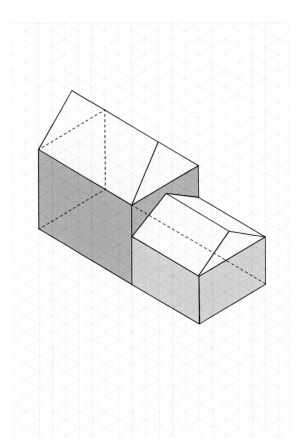

Step 2. Divide the volume into individual voids.

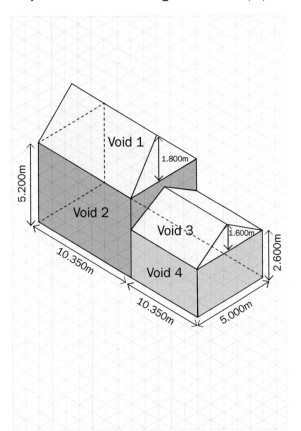

Step 3. Identify the different voids and place the dimensions on the sketch (once measured).

* Void 1 = ½ base × height × length
  = 2.500m × 1.800m × 10.350m = 46.575m³

* Void 2 = length × base × height
  = 10.350m × 5.000m × 5.200m = 269.100m³

* Void 3 = ½ base × height × length
  = 2.600m × 1.600m × 10.350m = 41.400m³ᵛ

* Void 4 = length × base × height
  = 10.350m × 5.000m × 2.600m = 134.550m³

TOTAL VOLUME   = 491,625m³

Step 4. Calculate the value of each void (above each other).

Step 5. Calculate the total value of all the voids to conclude the total volume (m³) of the building.

Area

This calculation is for understanding the size of any given area and is a vital formula to learn. The five-step plan below identifies how it can be achieved. Materials such as flooring and concrete can be accurately procured with as little waste as possible. This is achieved by breaking-up the floor plan into rectangles, working out their individual areas then adding up each area to conclude a total floor area.

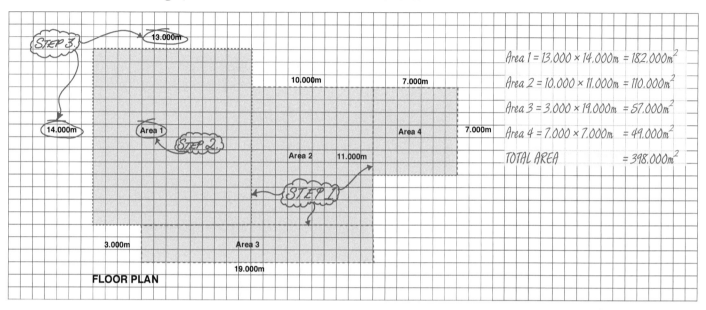

Perimeter of an irregular shape (i.e. floor plan): The same floor plan has been used in the example below to identify the perimeter of the same building. Once again, a simple five-point plan is used to identify the value of the boundary of the building line. Materials such as skirting board, coving and structural timber can be accurately procured with as little waste as possible using this formula. This method relies on the perimeter length being broken-up into shorter lengths or legs, measured individually and then added up to conclude an overall perimeter.

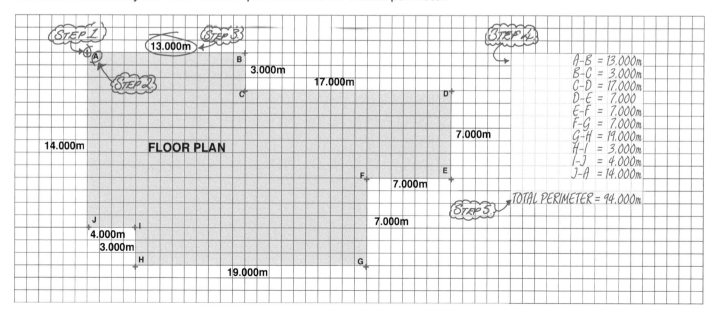

APPLICATION Recognise the shape and calculate the area of the materials.

1

Rectangle

Material: double doors

Shape: rectangle

Dimensions: 2.300m wide × 2.100m high

2

a

Trapezium

h

b

Material: slate roof

Shape: trapezium

Dimensions: a = 5.000m

h = 7.000m

b = 8.000m

3

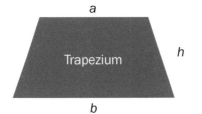

Parallelogram

h

l

Material: cement slate roof

Shape: parallelogram

Dimensions: l = 5.000m

h = 3.000m

4

h

Triangle

l

Material: masonry feature capping

Shape: triangle

Dimensions: l = 0.750m

h = 0.750m

5

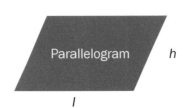

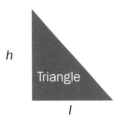

Circle

Material: metal alloy

Shape: circle

Dimension: 600mm diameter

KNOWLEDGE / Time

Within the construction industry the measurement of time is often defined in the contract documents. In broad terms, time is measured by one or more of the terms shown in the mind map below.

Working **day**
(× 8 hours excluding breaks).

Working **week** (× 5 days Monday to Friday and not including any bank holidays).

Working **hour** (60 minutes).

Measurements of time

Other: Often agreed between **the client and/or local planning authority**.

Such agreements may also refer to **out of hours working** (generally between the hours of 17:30 hrs to 07:30 hrs the following day).

Weekend working is also seen as outside normal working hours, although many contractors often work on Saturday mornings until midday.

Other hours are often referred to in the contract documents and sometimes have **limitations of working hours** clauses placed within them. This is to ensure that adjacent properties and/or businesses are not badly affected by the work.

REMEMBER

Such limitations of working hours clauses exist to ensure compliance with **local authority environmental health standards**. This means that the local community is not badly affected by excessive or lengthy exposure to residual risks such as noise, dust, vibration and congestion. **Such limitations help with governance and ensure sustainable levels of construction activity exist.**

KNOWLEDGE / Programmes

Within the construction industry, time and its associated application of labour, plant and materials are represented by both simple and complex schedules of activity. These schedules are referred to as **programmes**. Such programmes are not only restricted to the actual execution of work. Programmes can be used for procurement of materials, organising the design process, arranging deliveries and managing specialist trades or activities, along with almost anything else.

REMEMBER

Regardless of the size of the project, a programme exists to ensure that the interrelated tasks of any activity are logically sequenced and planned.

On larger, complex projects, such programmes have specialist programmers and managers to manage and control them and may have thousands of interrelated activities.

This includes any important health, safety and environmental factors that may affect the workers or production. Quality considerations are also often factored into programmes, and events such as hold-points are often planned. These hold-points give a chance for the team to confirm vital quality matters such as fire protection and structural connections can be checked and signed-off (witnessed) by specialist Engineers, Architects and building control officers. Important dates (i.e. approvals and handover dates), known as milestone dates, are also referred to.

KEY TERMS

Hold-point: A point during the design or construction process where the team pauses, checks and agrees the condition, progress or quality of an important aspect of the task at hand such as a vulnerable detail, an important part of a process or inspecting a building component that is about to be covered-up by the next activity.

Signed-off: A point at which the team agrees that the work constructed has reached an acceptable level of workmanship and complies with the specification. This is often recorded by the responsible person such as a site manager on a procedural form or electronically by email or digital acceptance.

Milestone dates: Dates in the programme or schedule that reflect important points in the design process and construction phase such as when the concept design is anticipated as being completed or when the building becomes structurally completed, watertight or ready for handover.

What does a programme look like and consist off?

The following example of a typical construction programme is a simple document that identifies the vital **preparatory tasks** (i.e. surveying the measurements of something or checking for hazards) **leading up to and including the actual task or activity** such as setting up a site. **Time is shown at the top** of the programme and **the activity at the left-hand side** of the document. Programmes are generated in digital or paper form. They can also be printed at any size and placed in the workplace so the workforce can read them and remain engaged with the construction process as it helps all stakeholders to work together. **Safety, quality and production are often improved by using programmes.**

Let's take a closer look.

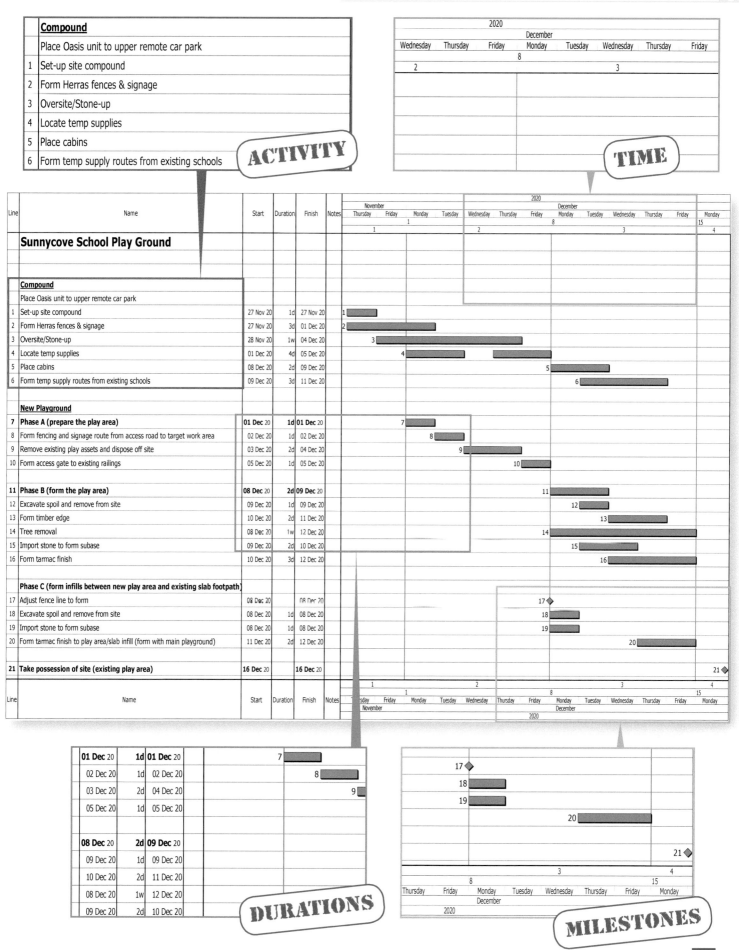

ACTIVITY

	Compound
	Place Oasis unit to upper remote car park
1	Set-up site compound
2	Form Herras fences & signage
3	Oversite/Stone-up
4	Locate temp supplies
5	Place cabins
6	Form temp supply routes from existing schools

TIME

2020							
December							
Wednesday	Thursday	Friday	Monday	Tuesday	Wednesday	Thursday	Friday
			8				
2				3			

Sunnycove School Play Ground

Line	Name	Start	Duration	Finish	Notes
	Compound				
	Place Oasis unit to upper remote car park				
1	Set-up site compound	27 Nov 20	1d	27 Nov 20	1
2	Form Herras fences & signage	27 Nov 20	3d	01 Dec 20	2
3	Oversite/Stone-up	28 Nov 20	1w	04 Dec 20	3
4	Locate temp supplies	01 Dec 20	4d	05 Dec 20	4
5	Place cabins	08 Dec 20	2d	09 Dec 20	5
6	Form temp supply routes from existing schools	09 Dec 20	3d	11 Dec 20	6
	New Playground				
7	**Phase A (prepare the play area)**	01 Dec 20	1d	01 Dec 20	7
8	Form fencing and signage route from access road to target work area	02 Dec 20	1d	02 Dec 20	8
9	Remove existing play assets and dispose off site	03 Dec 20	2d	04 Dec 20	9
10	Form access gate to existing railings	05 Dec 20	1d	05 Dec 20	10
11	**Phase B (form the play area)**	08 Dec 20	2d	09 Dec 20	11
12	Excavate spoil and remove from site	09 Dec 20	1d	09 Dec 20	12
13	Form timber edge	10 Dec 20	2d	11 Dec 20	13
14	Tree removal	08 Dec 20	1w	12 Dec 20	14
15	Import stone to form subase	09 Dec 20	2d	10 Dec 20	15
16	Form tarmac finish	10 Dec 20	3d	12 Dec 20	16
	Phase C (form infills between new play area and existing slab footpath)				
17	Adjust fence line to form	08 Dec 20		08 Dec 20	17 ◆
18	Excavate spoil and remove from site	08 Dec 20	1d	08 Dec 20	18
19	Import stone to form subase	08 Dec 20	1d	08 Dec 20	19
20	Form tarmac finish to play area/slab infill (form with main playground)	11 Dec 20	2d	12 Dec 20	20
21	**Take possession of site (existing play area)**	16 Dec 20		16 Dec 20	21 ◆

DURATIONS

01 Dec 20	**1d**	**01 Dec 20**	7	
02 Dec 20	1d	02 Dec 20	8	
03 Dec 20	2d	04 Dec 20	9	
05 Dec 20	1d	05 Dec 20		
08 Dec 20	**2d**	**09 Dec 20**		
09 Dec 20	1d	09 Dec 20		
10 Dec 20	2d	11 Dec 20		
08 Dec 20	1w	12 Dec 20		
09 Dec 20	2d	10 Dec 20		

MILESTONES

17 ◆	
18	
19	
20	
21 ◆	

'The Plasterer/Dryliner/Taper/ Jointer', https://www.youtube.com/ watch?v=lg53v-8l7zl.

Ratios

Materials such as plaster, mortar, concrete, resins and many other construction products rely upon being mixed both off site and on site. These production processes should be subject to codes of practice to ensure all the correct methods, equipment, materials and ratios are used during this vital period in the lifespan of the relative product.

Failure to adequately blend, mix or apply materials together under these strict conditions can easily result in compromising materials such as mortar for bonding masonry, structural concrete for load bearing or adhesion of plaster/render products on walls, to name but a few.

When some materials such as mortar or plaster are mixed, strict ratios associated with the production process must be followed to prevent such compromise.

What is a ratio?

In construction terms, a ratio is a like-for-like quantity of two or more materials that are intended to be mixed together.

Ratios should be referred to by a like-for-like number, volume or weight to ensure the intended portions are all of the same quantity.

'Lime Production – Why Learn Heritage Skills?', https://www. youtube.com/watch?v=Ve-m6gP3pZ8.

KEY TERM

Workability: The capability of a product when it is worked.

UNDERSTANDING The ratio of aggregates for a specified bricklaying mortar could be as follows:

- Mortar for a moderate mortar mix will be:

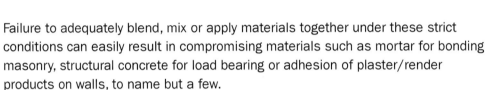

Cement : Lime : Sand = 1:1:5

The amount of water will be set at a maximum amount (to prevent the material failing once set) and a minimum amount to promote workability.

APPLICATION / What is the ratio of tools below?

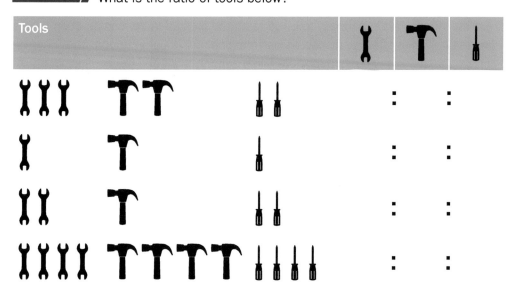

Tools						
🔧🔧🔧	TT	✎✎	:	:		
🔧	T	✎	:	:		
🔧🔧	T	✎✎	:	:		
🔧🔧🔧🔧	TTTT	✎✎✎✎	:	:		

KNOWLEDGE / Costs

Estimated costs help the construction industry to remain sustainable. In broad terms, the costs incurred by a contractor operating in the construction industry often must be recovered long (several weeks) after the expenditure by the contractor has occurred. This gives rise to commercial risk and many contractors have businesses that fail due to this reason alone – they simply have too many outgoings (see points 1–6 in the diagram at the bottom right of this page), for too long (point 7 in the diagram) and they are often not entitled to payment from the client until the work is completed.

Therefore, the estimated costs must be accurately scheduled by the contractor. Many contractors use commercial managers called Estimators for this very purpose. Estimators specialise in the discipline of predicting accurate construction project costs. They are not to be confused with the function of quantity surveying, as this is a slightly different, although similar, profession.

All the costs shown on the right and discussed on the following page will always have a time constraint (limit) in one form or another that will influence the estimated cost.

On completion of the project, the **actual costs** may vary, sometimes greatly, from the original **estimated cost**.

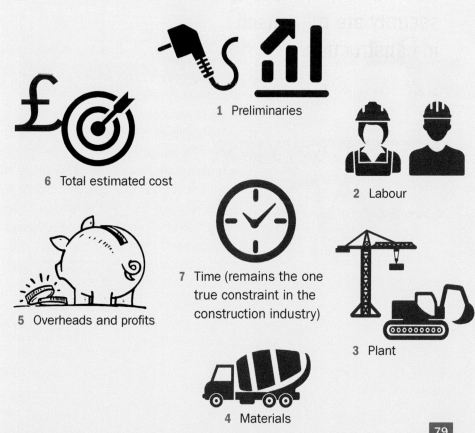

1 Preliminaries

6 Total estimated cost

2 Labour

7 Time (remains the one true constraint in the construction industry)

5 Overheads and profits

3 Plant

4 Materials

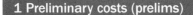

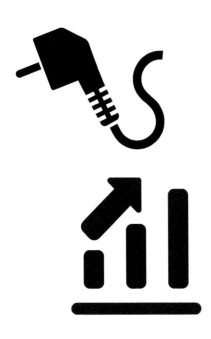

1 Preliminary costs (prelims)

Preliminary costs are normally referred to as the prelims in the construction industry and will often be issued as part the contract documentation. The prelims relate to any costs associated with administering the project, including:

- Site accommodation such as temporary buildings and welfare provision.
- Site-based staff such as managers, Engineers and bank persons.
- Fuel and energy provision during the construction phase.
- Bottled water, tea and coffee for meetings.
- Signs.
- Permits and permissions.
- Mobile telephones and digital devices.

Prelims are often added together to make an overall sum and this is multiplied by the total number of anticipated construction weeks (taken from the programme).

2 Labour costs

On larger projects, the labour provision is contracted to specialised sub-contractors and organisations such as concrete frame or bricklaying companies, which also provide materials and plant as part of their responsibilities. On smaller projects the labour is often directly employed by the company engaged to deliver the project. The costs for this element are valued on a trade-by-trade basis and will vary between disciplines. When there is a skills shortage, this cost can increase dramatically. Labour can be highly skilled and technically advanced, with some companies such as electrical providers requiring many weeks of notice prior to being appointed.

3 Plant costs

The cost of this element is often simply referred to as plant (not to be confused with roses or lavender) and relates to the costs associated with the provision of tools and equipment. These are often provided by the specialist contractors as part of their packages, given that they often must have specialised and expensive equipment, such as dumper trucks and cutting equipment, to complete their tasks. Smaller projects and organisations may refer to plant as the costs for purchasing smaller tools such as drills, hammers and handsaws. Crane hire is a cost that is familiar to both small and large organisations as they often both require heavy equipment and materials lifting into place. But to purchase and maintain a crane would be cost prohibitive.

4 Materials

Materials cost, as the name suggests, is the total cost for materials such as timber, fixings, board materials, plaster and decorative products. The quality of materials varies greatly, dependent on their cost and when and where they were manufactured. Modern materials used in the construction industry are manufactured to the highest standards, but nevertheless should be immediately checked on delivery to site as they can easily be damaged during transportation or while in storage. Many contractors provide their own materials for quality assurance reasons, to ensure they control and produce the very best quality possible. Larger construction projects, where large volumes of concrete or steel are to be used, may source their own materials to ensure that they do not encounter shortages.

5 Overheads and profits

Overheads: The types of overheads will generally be very similar regardless of the size of the organisation. Here, the essential costs such as running the business, office space, rates, energy consumption, advertising, marketing stationery and support staff costs are forecast and recorded. The scale of the overheads will vary dramatically though, dependent on numbers of employees, size and location of offices, as well as the size and type of projects that the company specialises in delivering.

Profits: This is the ultimate commercial objective of any viable and sustainable business. The profit of the business relates to the financial gain that is made by the business as a result of trading in the construction industry. Regardless of the size of the organisation, the forecast and actual profit relate directly to its success. Profit is essentially the difference between the actual total cost of producing a construction-related product or service and the amount a client has paid for that product or service.

6 Total estimated cost

The total estimated cost is a snap-shot in time of the anticipated cost of:

Preliminaries + Labour + Materials + Plant + Overheads and profit

The total estimated cost will be directly related to a period (**time**) that the contractor considers is feasible to deliver the project, often defined as 'weeks' and always within a contract with the client. **The total estimated cost is and remains a target not a certainty**, until the project is completed.

7 Time

The time element often remains a risk owned by the contractor so that provision for challenges during the construction process (within reason) such as inclement weather, poor health and safety and inefficient working practices are not rewarded.

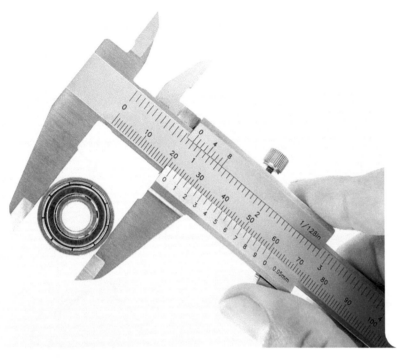

AC2.3 Set success criteria for completion of construction tasks

You need to know what you are reasonably capable of producing in a practical way. This means that you must set yourself some criteria that can be achieved when you present your construction products. Regardless of what discipline, craft or trade you are going to undertake, you must set yourself some quality targets and deliver them.

The three success criteria that you should set yourself are:

1 Levels of tolerance

2 Timescales

3 Quality.

Once you have set these criteria, you will have formed a quality-control-type document that can be referred to during the process and presented as part of the construction process on successful completion of the task. This document will help you to demonstrate you are focused on the delivery of quality production in a sustainable and health and safety conscious way.

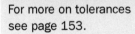

LINK

For more on tolerances see page 153.

SKILLS // Tolerances

Your skills to identify what are acceptable levels of tolerance will change during your construction careers as you become familiar with different types of processes and the application of the ability based on your talents and those of others too.

It is vital, though, that you set the highest standards that can be achieved within a reasonable timeframe to achieve acceptable and, if possible, superior quality. After all, if you don't then your time and resources will have been inefficiently used or even wasted. Once again, you have returned to the three success criteria:

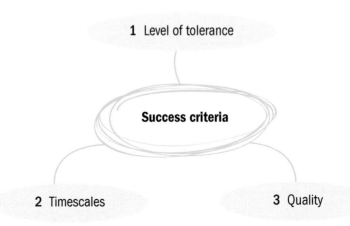

1 Level of tolerance

Success criteria

2 Timescales

3 Quality

KNOWLEDGE The following are the levels of tolerance for some of these refurbishment-related projects:

1 Textiles

- **Pelmets:** Textiles should be formed so that the texture, weave or pattern is parallel to adjacent surfaces behind the textile material to +/− 3mm.
- **Curtains:** Ditto.
- **Wall coverings:** Ditto.

2 Wood and carpentry

- **Hanging a door:** A door should be no more than 10mm out of plumb. The door should not be distorted more than 5mm across its width and 9mm across its height. The door should have a gap no greater than 5mm between the door and frame (top and sides) and 5–22mm between the bottom of the door and the floor.
- **Constructing a frame** (dummy window or door frame): The door frame should not appear distorted when fixed inside the opening. The frame should be square and no more than 5mm difference across the diagonal dimensions on the inside of the frame.
- **Attaching a skirting board:** The skirting should appear uniform when viewed at 2m and fixed at 450mm centres with single fixing if it is no greater than 75mm high. If it is greater than 75mm high, then the skirting board should be fixed with two fixings at an appropriate vertical distance apart.
- **Constructing a timber stud partition:** The timber stud should not appear distorted when fixed inside the opening. The stud should be square and no more than 5mm difference across the diagonal dimensions on the inside of the frame. The stud should be no more than 10mm out of plumb over its vertical height.

3 Brick

- **Use of and fixing wall connector ties or frame cramps:** Fixed at the first course and every sixth course.
- **Cutting bricks:** Visibly square to the edge or face of the brick.
- **Constructing a brick wall:** No higher than 1.000m × 5 courses long in a stretcher bond. A tolerance of +/− 6mm should be applied to all tolerance checks for this exercise.

4 Plaster and plasterboarding

- **Apply plasterboard.**
- **Apply skim coat to plasterboard:** Over any area of 300mm there should be no sharp deviance of more than 4mm. Over a 2m distance there should be no deviance of more than 5mm.
- Check what fire and sound details will be required around the head, sole and junctions of penetrations such as sockets and plugs.

'The Plasterer/Dryliner/Taper/Jointer', https://www.youtube.com/watch?v=lg53v-8l7zl.

5 Decoration

- **Apply emulsion to the wall** (as manufacturer's instructions): A uniform finish (when dry) should be achieved with no visible drips. Should have a solid colour,with a consistent tone. Any visible brush marks (minor) should follow a uniform direction.
- **Apply gloss paint to a panel door** (water-based gloss is acceptable): A uniform finish (when dry) should be achieved with no visible drips. Solid colour and tone should be visible, minor brush marks should follow a uniform direction.
- **Apply wallpaper or lining paper to a wall that has an internal corner and a light switch** (in whole or substantial part): The paper should be fully adhered to the surface with no visible air pockets or any debris visible through the paper.

6 Tiling

- **Apply tiles to a wall**: Tiles should appear consistent with one another and appear in a uniform bond. The grout should be solid with no evidence of spacers within. The tiles should be clean and smear free.
- **Apply tiles to a floor**: Tiles should appear consistent with one another and appear in a uniform bond. The grout should be solid with no evidence of spacers within. The tiles should be clean and smear free.
- **From a patch repair**: Tiles should appear consistent with one another and appear in a uniform bond following the line of the existing surface. The grout should be solid with no evidence of spacers within. The tiles should be clean and smear free. The grout should blend into that of the adjacent existing grout for a repair and not appear dissimilar.

7 Electrical

- **Add a new power socket** (services not live): The new socket should be wired following the manufacturer's instructions. The socket should be installed so that it is level, with the screw heads (if visible) following a consistent line.

8 Plumbing

- **Add a waste and taps to a sink unit** (services not live): The taps and waste should be fitted following the manufacturer's instructions as products vary greatly.

9 Heritage skills

- **Drystone walling** (replace or patch a section of damaged or missing drystone walling): The replacement 'face' stones should be of uniform size, shape and appearance (re-used if possible). 'Through' stones should be left in place when possible and/or repositioned with minimal movement/disturbance.
- **Roofing** (replace or patch a section of damaged or missing roof): The replacement roof materials should be of a uniform size and have the same appearance as the existing materials. The line of the new material should follow that of the existing materials, bond and pitch of the existing roof.

'Drystone Walling', https://www.youtube.com/channel/UCQSczLjBUv8ZWcf9mIuYv-Q/videos.

KNOWLEDGE / ## Timescales

Adequate time must be given to the essential preparatory and post-construction tasks known as 'housekeeping'. Timescale is best understood by breaking down the process into some simple but fundamental activities. It should be no surprise that the actual execution of the task is only part of this process. The task (in whatever form) should be approached giving adequate time for:

- Reading and referring to the drawings, specification and schedules.
- Preparing and researching your approved risk assessments and method statements. Consider the provision of COSHH sheets where applicable (e.g. 'Working with mortar, plasters or adhesive').
- Procurement and distribution of materials.
- Preparation of the target work area, taking into consideration how your proposed work will affect others working adjacent to you and vice versa.
- The planned duration of the work/task itself.
- Clean-up and protection (housekeeping) of the work and its adjacent areas, including the provision of barriers and signs to prevent damage to your's and other's work.

The use of a simple programme is advised here and is referred to in AC2.2 of this unit.

REMEMBER

Draft a simple a programme and keep it with you while you work, as this can also act as a helpful checklist.

LINK

For programmes see pages 75–77.

KNOWLEDGE / ## Quality

Achieving acceptable quality within the construction industry can be challenging when working with so many different stakeholders, each of whom has different agendas and motivations to complete their own task. Working in and with the elements is also challenging, particularly in winter. That said, to deliver a high standard of finish is equally as rewarding and can give great satisfaction to individuals and teams alike.

In the construction industry the recognised institutional standard for quality is currently referred to as ISO 9001. The primary benefits of working towards this level of quality attainment help both the client and contractor achieve their obligations under the contract by working to a process of checking and improving to a set of pre-determined objectives. This process is continually improved on a project-by-project basis.

UNDERSTANDING / There are eight principles to this process:

1 Focus on the customer and achievement of their expectations.
2 Clearly defined leadership to help guide the delivery team to success.
3 Inclusivity of everyone who is involved in delivering the project.
4 Installation of processes that focus on improvements and marginal gains.
5 Systematic management approach to processes that have been installed.
6 Continual improvement is attained through transparent (easily understandable) working, and mechanisms best practice is recognised and promoted.
7 Prior knowledge and experience are recognised as a conclusive means of factual-based decision making.
8 Supply chain relationships are encouraged and promoted to ensure seamless working wherever possible.

REMEMBER

This approach to quality is not just focused on outputs alone (i.e. producing a superior finished masonry wall or roof), it's all about the process and the journey to achieving quality as opposed to the destination of a great quality product. It is often referred to as a 'quality culture' or 'quality behavioural tool'.

REMEMBER

The **three-point** process of **preparation, execution** and **cleaning-up** makes all activities **safer, more efficient and therefore more sustainable**. If you follow this simple process every time you work, then your skills and reputations are likely to improve and you will become **contractors of choice** and your **sustainability** becomes even greater.

KEY TERM

Dry-bond: The process of offering-up a material and placing it in-situ (in place) and assessing its suitability, prior to bonding it in place permanently.

AC2.4 Prepare for construction tasks

Within the construction industry, once you start the activity of preparing to execute the task, you are starting the task. The preparatory activities before construction of a masonry wall, for example, are as important as the construction process of building it. Similarly, follow-on cleaning-up of the adjacent area and tools is equally a part of the process. It has already been highlighted that if you 'fail to plan, you are planning to fail'. Now is the time to put the theory into practice.

SKILLS / **Preparing materials**

Let's look at what you need to do and how you do it when you prepare materials and start the production process. This is the moment of **realisation**.

✓ **Check materials for defects**   ✓ **Cut materials**

✓ **Organise materials**   ✓ **Set out materials**

✓ **Measure materials**   ✓ **Dry-bond materials**

✓ **Mark-out materials**   ✓ **Mix-mortar materials**

KNOWLEDGE / Construction terminology is used to refer to how methods and materials are realised in the workplace. This reference relates to how accurate the materials have been designed, manufactured, and formed or constructed. Let's start using this terminology when referring to realisation or working. When reading this terminology remember that you will be functioning in a real 3D environment, so always consider that just because a design or concept works in 2D it may not always be the case it will work in 3D (i.e. reality). When geometry of materials and workmanship is consistent and communicated with this terminology effectivity, accuracy and quality will naturally follow.

What is the difference between the D's?

- **1D** is the difference between a point on a line, length or distance, or near or far. In the UK construction industry, this is measured in metres.

- **2D** is geometry that has length, width and area. In the UK construction industry, it is measured in metres^2.

- **3D** is geometry that has length, width, depth and volume. In the UK construction industry, it is measured in metres^3.

See how the simple length of a 1D line becomes 2D when the area is defined with the inclusion of width. 3D development is rapidly achieved by adding depth to achieve volume.

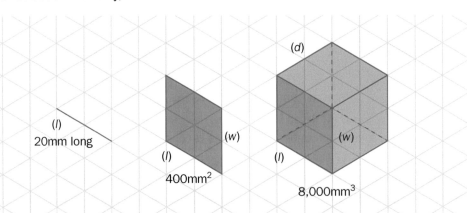

(l)
20mm long

(d)

(w)

(l)
400mm^2

(l)

(w)

(l)
8,000mm^3

UNDERSTANDING Terminology commonly used in the construction industry varies between regions and sometimes occupations, but the following examples are commonly used and can be applied to all skills, crafts, trades and disciplines. Unless by exception (an exceptionally non-uniform shape), the whole industry aspires to be as **true** and reflective to these parameters as possible. The industry uses the following terminology as various kinds of **destinations** (e.g. the height of the windowsill board is 950mm above finished floor level), where working on the same project (building a house, for example). **This gives each trade a definitive horizontal or vertical reference point from which to work towards or away from.**

In the following table, the colours refer to:
A detail that is formed (i.e. an Arris).
Used for setting out and checking.
A reference point.
An aspiration or acceptable standard.

The universal symbol for a right angle (90°).

KEY TERM

Parameters: requirements or measurable aspiration.

Term	Symbol	Definition	Example
Arris	A	The external corner that is formed where two surfaces intersect. This can be any material but is generally referred to when describing masonry and timber materials.	*Lyn the carpenter has removed the arris from the door lining in the kitchen as it will minimise the risk of children hurting their fingers on the sharp edges of the planed timber.*
Centre line	CL	The centre line is the line that is referred to when setting-out materials in relation to a building or room. It can also refer to the actual centre-line of a material of component.	*The ceramic tiles were positioned on the sill so that they were either side of the centre line, this lined in with the centre line of the new sink.*
Consistent	C	(Consistency) is a word generally used as a positive symptom of regularity and uniformity in the construction industry. However, it can also be used in a negative context.	*The consistency of the gloss paintwork to all the internal doors was first class.* *The approach to health and safety by John was consistently poor.*
Datum	D	This is a fixed point of a scale. On projects where the building is to be newly constructed the datum can be found on Ordnance Survey maps and relates to a fixed horizontal point above sea level. This point will remain the same through the project, so the workforce has a consistent point to refer to. On refurbishment projects, the datum is often set by the site manager and is frequently 1m above a fixed point (a special stud or mark is used to determine this point) on the finished floor.	*The datum for the project is set to the Ordinance Survey benchmark in the street; it has a value of 22.600m above sea level. The new house has a proposed finish floor level of 22.800m, so it will be 200mm higher that the datum point.* *The datum point for the refurbishment of the kitchen at Sunnycove Lodge is the top of the ceramic tiles on the windowsill, in the kitchen.*

Term	Symbol	Definition	Example
Dimensional coordination		This is a design concept that allows different materials to be manufactured to compatible and therefore more efficient sizes and shapes. In the construction industry, a 600mm suspended ceiling tile and grid system is commonly used in office design together with a 600mm carpet tile. The rest of the office is then planned around this concept.	*Sara the Architect decided to only use an American dimensional coordination scale as she wanted the overall room and its content to be dynamic and greater in scale than the other houses on the street. Sara selected 600mm wide tiles and worktops for the kitchen to achieve dimension coordination.*
Grid		The grid is another design tool that the whole team can utilise when designing, setting out or constructing with a variety of materials. This is directly associated with dimensional coordination and the datum, as they often all interact with one another in the completed overall effect of the room or building project.	*All the products in the kitchen have been specified to be compatible with a 600mm grid. This will give a consistent visual feel to the tiles, window, worktops and flooring.*
Finished floor level		This is the actual level that the users of the house will be standing on when completed. When working on exiting properties where the floor is to remain, it is often used as the datum during the works, as it remains constant and is already in place.	*The site manager, Lyn, instructed me to use finished floor level as the datum in the kitchen so that the dishwasher will fit into the space under the worktop next to the sink perfectly.*
Head (not to be confused with a reveal)		The head section of a window frame or door frame is referred to considerably in the construction as it sets a horizontal reference point within a room or when viewing a building's elevation externally. If window and door heads do not align horizontally when placed above each other the effect can look inconsistent and not good throughout.	*Elin the Architect decided that the window head of the window in the kitchen needed to be at least 200mm higher than the head of the adjacent door to ensure the view out of the window when working at the sink was maximised.*
Jamb (again, not to be confused with a reveal)		The jamb is the vertical section of a window or door (often at 90° to the head).	*Elin the Architect decided to design wide window jambs to give the impression of robust and secure windows and doors. When completed, the windows were proportional to the reveals and the margins were consistent.*

Term	Symbol	Definition	Example
Line	L	A general term that relates to the 1D, 2D or 3D straightness of a line of components, products or materials.	*The stretcher bonded wall lined in well with the external walls of the main house. Or, Although the plaster patch on the existing wall was over two metres wide and two metres high the plasterer did a great job patching it. It lined in perfectly. Or, Use the line of the existing skirting board on the right-hand side of the wall to set out the new timber stud partition.*
Level	Level	A very overused term in the construction industry. It is often misused for just about anything and everything related to materials when they have been fixed. This word means: An engineered tool device for establishing a horizontal line or plane usually by means of a bubble in a liquid that shows adjustment to the horizontal by movement to the centre of a slightly bowed glass tube. The same device is often used to establish plumb/vertical alignment.	
Nose (or nosing)		This is the foremost protruding part of a stair tread.	*The stair nosing was made from hardwood, so it was very hard wearing.*
Margin(s)	=M=	The distance by which one material is different from another. It is often referred to just before plastering walls and fixing skirting boards and architrave.	*The margins around the window frame were all consistent. Or, The margins where the architrave met the door lining were all the same, the quality of the carpenter's work is fantastic.*
Plumb	P	This term is taken from an ancient device (plumb bob) that was and still is used to determine a true vertical line.	*The verticality of the reveal was checked by the surveyor and he concluded that it was truly plumb.*
Reveal	R	The reveal is the area where there is a recess in a wall. The reveal is the portion of the recess that has the less depth compared with the surround walls or surface. It is the surface area either side of an aperture.	*The reveals to the kitchen window were both equal.*

Term	Symbol	Definition	Example
Square	S	This term is another word that is often misused. In the construction industry it should only relate to the accuracy of a 90° angle.	*The kitchen worktop was well made and was proved to be square when I checked it with my angle-finder device. Or, The kitchen work top was truly square, it was a rectangle with four equal sides measuring 600mm each and its diagonals were of equal length to one another, they both measured 848mm.*
True	T	This is a term often used in the construction industry and relates to the material as being compliant to the specification, drawing, design briefs and/or schedules.	*Iwobi the banksman checked the skirting board delivery for quality prior to being unloaded. Iwobi found the 5m lengths of 75mm × 18mm hardwood pencil-round skirting to be true. Iwobi approved the delivery and it was incorporated into the kitchen base units.*
Windowsill level (internal)	WSL	The finished level of the internal windowsill inside the room or building. This is an important height as it often determines the glazing specification of the window within any fixed furniture such as kitchen units.	*Alun Perry, the interior designer, wanted to have a solid band of ceramic tiling above the sink to provide a splashback below the windowsill that was easy to clean.*
Windowsill level (external)	WSL	The finished level of the internal windowsill outside the room or building. It is often lower that the inside sill to help prevent water ingress below the window frame.	*The stretcher bonded brickwork below the external concrete windowsill was stained by green algy.*

APPLICATION / Let's take time to focus on the materials that you can work with to demonstrate your skills, knowledge and understanding. By applying the principles, you are prepared to realise the projects available to you. Use the following image of the newly located kitchen sink unit at Sunnycove Lodge and refer to the terminology to help you understand how to form constant details.

Let's see what this looks like.

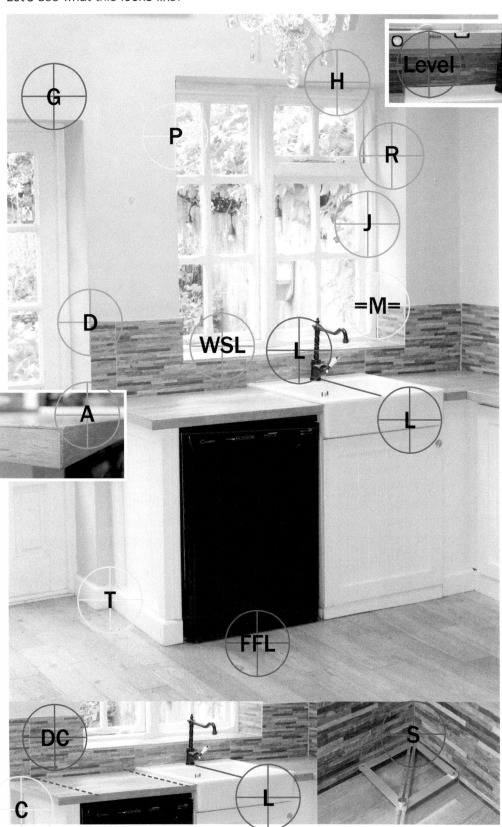

Sunnycove Lodge kitchen under construction.

'Brickskills, the Build: A Step-by-step Guide to Sunnycove Lodge', https://www.youtube.com/channel/UCQSczLjBUv8ZWcf9mIuYv-Q/videos.

LO3 Be able to use construction processes in completion of construction tasks

AC3.1 Apply techniques in completion of construction tasks

TEXTILES

① VISUAL CHECKS FOR DEFECTS:

- Check consistency of pattern.
- Check for faded textiles. Have they been left in the sun?
- Grease marks?
- Oil marks?
- Is the material creased too much (beyond ironing)?

② DIMENSIONAL CHECKS FOR ACCURACY:

- Textiles should have consistent and uniform weave/texture.
- Textiles should have a consistent pattern/symmetry.
- Consider if the pattern or weave can be joined.

③ ORGANISING MATERIALS FOR SAFETY AND PRODUCTIVITY:

- Textiles should always be stored on a roll or folded into large squares.
- Always store textiles out of direct sunlight.
- Always inspect the full length of material when it is removed from stores or purchased.
- Textiles should be stored off the floor and out of any damp draughts.
- Textiles should not be exposed to damp environments.

KEY TERM

RAMS: Risk assessment and methods statement; term used in the construction industry referring to two key documents that are needed to be approved by the site manager prior to starting work.

④ SETTING OUT TO ENSURE ACCURACY AND QUALITY:

- Always work the fabric on large work-surfaces that are clear of other materials.
- Use sharp fabric-type tailor's chalks for marking and setting out.
- Ensure your fabric is free of creases and bubbles/waves/ripples before you measure.
- Use a solid fixed rule where possible for measuring and marking.
- Use a flexible tape as a secondary means to measure textiles.
- Always check the fixing capabilities of your chosen textiles; consider if they can be adhered/glued/pinned or stapled prior to selection.

TOP MUST DO'S

✓ Always follow your cutting RAMS.

✓ Use clean tools and cutters.

✓ Check your textile up-close and from afar.

✓ Select a low maintenance material that does not require specialist cleaning.

5 MEASURING TO ENSURE ACCURACY:

- Keep it simple – don't attempt complex shapes if you are not ready.
- Select a repeater pattern if you don't use plain textiles.
- Understand the scale of the pattern in relation to the proposed location of your work.

6 MARKING TO ENSURE ACCURACY:

- Measure twice and cut only once.
- Use a chalk wheel or chalk pen.
- Consider using an erasable biro-type pen.
- Use a template if the shape is not rectangular.
- Pin your textiles prior to cutting to help stabilise the fabric.
- Consider using disappearing ink – it does just that over time.
- Consider using temporary threads (tailor's tacks) to help you mark-out, if you are unable to mark the material.
- Consider using template paper for complex shapes.

7 CUTTING FOR SAFETY AND ACCURACY:

✂ Consider what tool is appropriate for the textile that you are working with; some materials are designed to be robust and are cut/scrag resistant.

✂ Practice how the cut works on spare pieces or off-cuts first.

✂ Use sharp tools, as blunt cutting equipment will scar the textile.

MUST HAVE'S

✓ The correct PPE.
✓ A true back material to fix to.
✓ The back-board must be capable of being fixed to.
✓ Clean damp cloth at hand when working to clean-up.

WHAT THE EXPERTS WILL BE LOOKING FOR:

1 Alignment of pattern/weave with the back-board of the leading edge of the product.
2 Fixings should not be visible through the fabric.

KEY TERMS

Template paper: a prepared document that can be used as an aid when delivering a task.

Off-cut: The waste material left over from the process of cutting a material such as wood or plasterboard.

WOOD

① VISUAL CHECKS FOR DEFECTS:

- Straightness over the full length of the product.
- Aris is still intact and not damaged.
  - A. Knots
  - B. Bows
  - C. Shakes
  - D. Cupping
  - E. Mildew
  - F. Rotting
  - G. Splits
  - H. Twists.
- Insect infestation.
- Correct species has been delivered.

② DIMENSIONAL CHECKS FOR ACCURACY:

- Check cross-sectional width.
- Check cross-sectional depth.
- Check overall length is adequate for the purpose.
- Moisture content: is the timber likely to need more drying time?

③ ORGANISING MATERIALS FOR SAFETY AND PRODUCTIVITY:

- Keep materials shrink-wrapped until they are used.
- Store flat if possible or vertically if banded together, and restrain from falling.
- Store like species together.
- Always keep off-cuts for practising.
- Handle with care – timber is fragile and can bruise easily.

④ SETTING OUT TO ENSURE ACCURACY AND QUALITY:

- Correcting centres for primary fixing of the product.
- Correct number of fixings.
- Location of the product in the aperture (i.e. door or window).
- Consider any other materials that may need to be set out first (adjacent floor finishes).
- Check moisture content (MC) – if internal use timber is less than 13% MC.
- Has the timber been kiln-dried?

KEY TERM

Cupping: The natural, localised bending across the cross-section of timber usually as it dries out.

TOP MUST DO'S

- ✓ Always follow your RAMS.
- ✓ Follow your supervisor's directions.
- ✓ Cut in a ventilated area.
- ✓ Practise cutting on off-cuts.
- ✓ Secure your materials with a cramp or vice when cutting.
- ✓ Wear eye protection when cutting.
- ✓ Wear the correct gloves when handling and cutting timber.

5 MEASURING TO ENSURE ACCURACY:

- Measure twice and cut only once.
- Use a *calibrated* tape that is in good condition.
- Ensure that the *graticules* on the tape can be easily read.

 6 MARKING TO ENSURE ACCURACY:

- Always use a sharp HB pencil.
- Use a pencil sharpener to keep your pencil very sharp.
- Use a set square to set out your cross-cut marks.
- Ensure your marking-out for cross-cutting is placed on all sides of the material.
- Use a sharp saw – recycle blunt saws.

KEY TERMS

Calibrated: The successful process of checking the measurement of an instrument.

Graticules: A defined set of consistent horizontal and/or vertical lines that help define measurement.

MUST HAVE'S

✓ PPE
✓ Sharp saw
✓ Claw hammer
✓ Tape measure (5m)
✓ Pencils (and sharpener)
✓ Set square
✓ Three-piece set of chisels
✓ A5 notebook
✓ Block plain
✓ Sanding block
✓ Screwdriver set.

7 CUTTING FOR SAFETY AND ACCURACY:

- ✂ When using a handsaw, use long confident strokes.
- ✂ Do not apply too much downward pressure to the saw.
- ✂ Always secure the material that you are cutting with a cramp or vice.
- ✂ Consider where and how waste material will fall after cutting. Will it splinter off?
- ✂ Ensure that you always use adequate cut-resistant gloves of the correct level/type.

WHAT THE EXPERTS WILL BE LOOKING FOR:

1 Compliance with tolerances (no large gaps).
2 Plumb vertical work.
3 Straight and level horizontal work.
4 No splinters and smooth finish.

BRICKWORK

① VISUAL CHECKS FOR DEFECTS:

- Check consistency of brick shape.
- Check consistency of brick colours.
- Check arris of brick for chips (called pearling).
- Check for evidence of poor storage such as efflorescence/salts.

② DIMENSIONAL CHECKS FOR ACCURACY:

- Bricks should be consistent in size and shape (nominally 65mm × 102mm × 215mm).

'Types of Brick', https://www.youtube.com/watch?v=QKEruGU3ZK4.

③ ORGANISING MATERIALS FOR SAFETY AND PRODUCTIVITY:

- Always mix the same type of bricks from different packs so that they contrast with one another.
- Bricks should be delivered on pallets and shrink-wrapped.
- Have adequate materials ready at hand.
- Always protect your bricks and work from the elements by covering them up overnight.

TOP MUST NOT DO'S

✗ Do not over-handle materials.
✗ Never form brickwork in wet or frosty conditions (+3 degrees and rising).
✗ Do not fail to prepare.
✗ Never throw bricks.

TOP MUST DO'S

✓ Always wear PPE.
✓ Always use dry bricks.
✓ Always have adequate materials at hand.
✓ Always check the setting out of the first course.
✓ Always clean your tools at the end of the day.

④ SETTING OUT TO ENSURE ACCURACY AND QUALITY:

- Always place a string line to the arris of bricks on the top course.
- Dry-bond bricks when practising your chosen bond.
- Do not rush your work.
- Barrier off your work area.
- Regularly check your line is not loose or snagged (people often trip over these on site).

REMEMBER

When mortar is mixed a chemical reaction occurs which can be harmful. Always wear gloves and clean your tools regularly. Mortar and cement can burn!

MEASURING TO ENSURE ACCURACY:

- 5m tape measure
- Set square
- Gauge rod
- Clean measuring tool
- Understand where the wall ties and support are to be fixed.

 MARKING TO ENSURE ACCURACY:

- Brick pencil
- Chalk line
- Chalk or crayon to mark floor
- Measure twice and cut once.

7 CUTTING FOR SAFETY AND ACCURACY:

- ✂ RAMS.
- ✂ Wear full PPE.
- ✂ Remember that bricks often shatter when being cut.
- ✂ Check that everyone is wearing PPE in adjacent areas.
- ✂ Use a bolster fitted with a shock-absorbing guard.
- ✂ Strike the face of a brick once when cutting to prevent multiple fracturing.

MUST HAVE'S

- ✓ Trowel
- ✓ Hammer
- ✓ Bolster
- ✓ Pointing tool
- ✓ String line and pins
- ✓ Spirit level
- ✓ Buckets
- ✓ Hand brush and clean-up buckets
- ✓ Full perp and bed joints.

KEY TERMS

Perp: relates to the perpendicular mortar joint that is present on all masonry. This is the shorter vertical joint that is often finished in different ways depending on the specification.

'How to Cut Bricks', https://www.youtube.com/watch?v=gCN68N3Djxo.

WHAT THE EXPERTS WILL BE LOOKING FOR:

1. Clean and consistent brickwork.
2. Uniform *perp* and bed joints.
3. Work that is constructed to a line.
4. Plumb work.

PLASTER AND PLASTERBOARD

'The Plasterer/Dryliner/Taper/Jointer', https://www.youtube.com/watch?v=Ig53v-8I7zI.

① VISUAL CHECKS FOR DEFECTS:

- Ensure the surface to be plastered is smooth and dust free.
- Check if there are adequate fixings in place.
- Check your grounds/guides are secure.
- Completed plaster surfaces should be smooth and free of crazing.
- Have the right types of beads or angles been selected?

② DIMENSIONAL CHECKS FOR ACCURACY:

- Agree the scope of work before you start.
- Understand the finished area that will be visible.
- Are expansion/cracking joints required?
- If studwork is used, check that the studs are there by looking for fixings.
- Check window and door reveals are consistent.
- Check the use-by date on all materials.

③ ORGANISING MATERIALS FOR SAFETY AND PRODUCTIVITY:

- Don't mix too much material; check the batching.
- Ensure plasterboard is staked flat.
- Never lean plasterboard against a wall, as it will likely slip and injure the workforce.

④ SETTING OUT TO ENSURE ACCURACY AND QUALITY:

- Double-check the line and level of your grounds/guides.
- Have all joints been reinforced with tape or scrim?
- Have the guides been placed consistently?

KEY TERMS

Crazing: The random and irregular pattern often created when cementitious materials (things made from cement) dry out and crack in an uncontrolled fashion.

Expansion/cracking joints: Joints intentionally formed in concrete to control the location and extent of the natural tendency of concrete material to crack.

TOP MUST NOT DO'S

✗ Never lean plasterboard against a wall as it will likely slip and injure the workforce.

✗ Never use wet or damaged plasterboard.

✗ Never use plaster products externally.

⑤ MEASURING TO ENSURE ACCURACY:

- Check plasterboards are a consistent size and thickness on delivery.

⑥ MARKING TO ENSURE ACCURACY:

- Studwork should be lined and levelled prior to board application.
- Check if any timber back-boards are required to receive joinery or sinks at a later date.

7 CUTTING FOR SAFETY AND ACCURACY:

- ✂ Always wear the correct PPE when plastering or boarding.
- ✂ Cut-resistant gloves are essential when scoring or cutting.
- ✂ Eye protection is essential when cutting or mixing plaster materials.
- ✂ A timber saw can be used for cutting thicker plasterboards and sheet materials but can also make lots of dust. Use the correct type of dust mask when cutting.

TOP MUST DO
✓ Ensure plasterboard is staked flat!

'Plastering and Skimming', https://www.youtube.com/watch?v=wAPJm3rfScs.

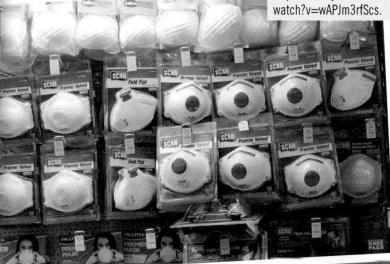

MUST DO'S
✓ Mix plaster material in a well-ventilated area (or outdoors).
✓ Plasterboard should be stacked flat, not vertically.
✓ Have a clean and readily available water source.
✓ Clean-up tools must be at hand.

WHAT THE EXPERTS WILL BE LOOKING FOR:

1 Smooth and consistent finish without crazing.
2 Equal margins around windows and door reveals.
3 Plum walls and 'true' lines at junctions of other material such as skirtings and architraves.
4 All internal and external corners should be formed at 90 degrees unless otherwise specified.

DECORATION

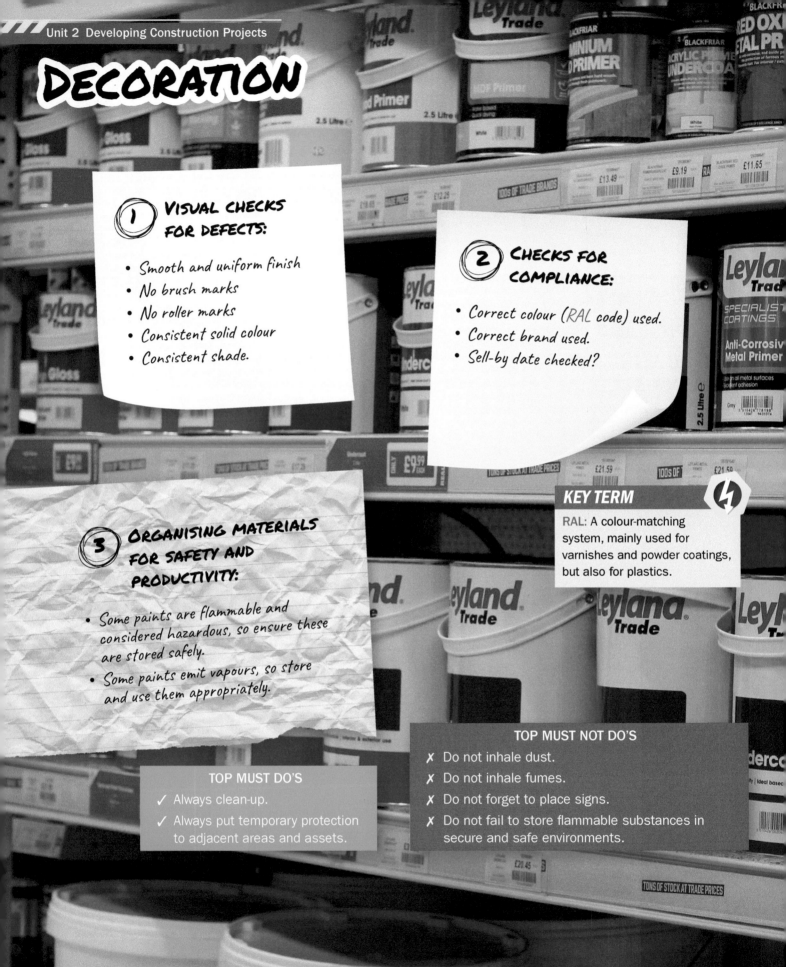

1 VISUAL CHECKS FOR DEFECTS:

- Smooth and uniform finish
- No brush marks
- No roller marks
- Consistent solid colour
- Consistent shade.

2 CHECKS FOR COMPLIANCE:

- Correct colour (RAL code) used.
- Correct brand used.
- Sell-by date checked?

3 ORGANISING MATERIALS FOR SAFETY AND PRODUCTIVITY:

- Some paints are flammable and considered hazardous, so ensure these are stored safely.
- Some paints emit vapours, so store and use them appropriately.

KEY TERM

RAL: A colour-matching system, mainly used for varnishes and powder coatings, but also for plastics.

TOP MUST DO'S

✓ Always clean-up.
✓ Always put temporary protection to adjacent areas and assets.

TOP MUST NOT DO'S

✗ Do not inhale dust.
✗ Do not inhale fumes.
✗ Do not forget to place signs.
✗ Do not fail to store flammable substances in secure and safe environments.

4 CHECKING TO ENSURE ACCURACY:

- Check the coverage each type of paint is expected to achieve.
- Products and brands vary in colour, so check the same ones are used.
- Scan existing paint colours whenever attempting to match colours.

5 PREPARATION FOR SAFETY AND ACCURACY:

▲ Ensure the correct grade sandpapers are used to prevent scaring.

▲ Only sand-down in well ventilated and controlled environments.

▲ Check the source of light while painting is sufficient.

MUST HAVE'S

✓ Spare rags for cleaning-up
✓ Temporary protection and dust sheets
✓ Masking tape
✓ Full set of clean brushes and rollers
✓ Brush cleaners (spirt or water based)
✓ Test area
✓ 'Wet paint' signs.

WHAT THE EXPERTS WILL BE LOOKING FOR:

1 Consistent colour, shade and texture of paint.
2 Consistent 'true' and straight 'cutting-in'.
3 Consistent matt, silk or gloss finish.
4 Cleanliness.

101

TILING

① VISUAL CHECKS FOR DEFECTS:

- Substrate must be dust/debris free.
- Substrate should not be freshly painted.
- Substrate must be clean.
- Substrate must be dry.

TOP MUST DO'S

✓ Always check the setting-out point.

✓ Always wear eye protection.

✓ Always clean-up as soon as you complete for the day, as the grout or adhesive sets rapidly and is difficult to remove when hard.

② DIMENSIONAL CHECKS FOR ACCURACY:

- Walls must be straight and true.
- Report any line, level or angle irregularities to the site manager of client prior to application.
- Check that internal and external corners are true.
- Consider if the walls are suitable for the size of tile selected.

TOP MUST NOT DO

✗ Never cut towards yourself.

③ ORGANISING MATERIALS FOR SAFETY AND PRODUCTIVITY:

- Check for broken tiles as soon as they are delivered or collected.
- Store your tiles away from walkways and in a secure area so they don't get hit or stolen.
- Check that all the tile manufacturing boxes have the correct brand/type of tile, as they often vary in shade or colour.
- Make sure that you have at least 5% more tiles than you need to cover waste and breakage.
- Only stack out the tiles you need for the period you are working, so that they are not exposed to theft or damage.
- Tiles are sharp, so be mindful of this.
- Ensure that the correct adhesive is specified, as different substrates such as timber floors and shower rooms will need different approaches.

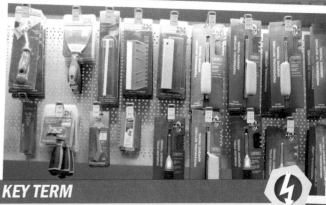

KEY TERM

Substrate: The material that is protected by and located below the surface material.

4 SETTING OUT TO ENSURE ACCURACY AND QUALITY:

- Always agree your setting-out point with the site manager or client before you start tiling.
- Draw the agreed datum for vertical and horizontal setting out on the substrate prior to starting tiling and show the client or site manager what the grid will look like.
- Take your time setting out and think if the setting-out point should be the centre line of the wall, from one corner to another, work from a window or feature or have a specially agreed point.
- Consider how skirting boards, coving or boxing-in will affect the final look and feel of the completed wall.
- Remember: measure twice and cut once!

5 MEASURING TO ENSURE ACCURACY:

- Check that the tiles are consistent in size and shape.
- Check that the tiles are not bowed or kinked by placing them on a flat surface to see if they rock.
- Always keep your off-cuts.
- Think if it is possible to get several cuts from one tile before you start cutting.
- Dry-bond where you can.

6 MARKING TO ENSURE ACCURACY:

- Use a mechanical pencil for greater accuracy.
- Use a china-graph pencil for very polished tiles.
- Use a metal straight-edge ruler or bevel instead of a tape measure if you can.

KEY TERMS

China-graph pencil: A special pencil that can write on smooth surfaces as well as paper and remains water resistant when wet.

Bevel: A guide that can be altered to form a desired and consistent angle, often used when working with wood and metal.

WHAT THE EXPERTS WILL BE LOOKING FOR:

1. Consistent colour, shade and texture of tiles.
2. Consistent true and straight cuts.
3. Consistent joint spacing.
4. No visible spacers.
5. Consistent grout surface.
6. Cleanliness of tiles is crucial, no grout clouds should remain on surface. Remember to dry the tile surface after washing down.

MUST HAVE'S

✓ PPE
✓ Metal measuring rule or bevel
✓ Tile cutter
✓ Pencil or china-graph
✓ Ceramic coping saw (for curves)
✓ Clean sponge and clean water
✓ Good quality adhesive applicator.

HERITAGE SKILLS: DRY BONDING TECHNIQUE OF MOST MATERIALS

'Heritage Skills Tutorial', https://www.youtube.com/watch?v=VO51SLV4e7o.

Dry bonding is the technique that can be applied to most heritage skills. This involves the systematic selections and offering-up of the materials that you are working with in an attempt to find suitably sized and shaped materials for your task. This method of working is favoured by many craftspeople and tradespeople because it allows you to visualise the best materials to select and use prior to committing to using them. It is also favoured because many heritage products are naturally sourced and differ greatly in size, shape and thickness. This method is time-consuming, so it takes longer to produce and complete tasks but it is a highly successful and ancient technique that produces a consistent finish, and reduces the need for mechanical cutting while reducing waste.

TOP MUST NOT DO

✗ Working with heritage skills, techniques, tools and materials can be incredibly rewarding. The environment too can often be relaxed and rural in nature. Do not be complacent to the hazards and associated risks when working, as all construction projects can be potentially hazardous.

TOP MUST DO's

✓ Always make sure your supervisor knows what you are doing.

✓ Wear the correct PPE while following your RAMS.

1   The image below shows three selected stones that have been dry bonded and are in the process of being moved to their final location (represented by the red dashed line) so that they form the best possible structural bond.

2   See how the courses below have been jointed-up with their perp and bed joints formed progressively.

3   See how the masons have started with a levelling-course to make the wall as consistent as possible.

WHAT THE EXPERTS WILL BE LOOKING FOR:

1   Consistency of bonding natural materials together.

2   As natural a finish as possible but with consistently selected materials that are similar in size and shape and are logically placed in order.

AC3.2 Apply health and safety practices in completion of construction tasks

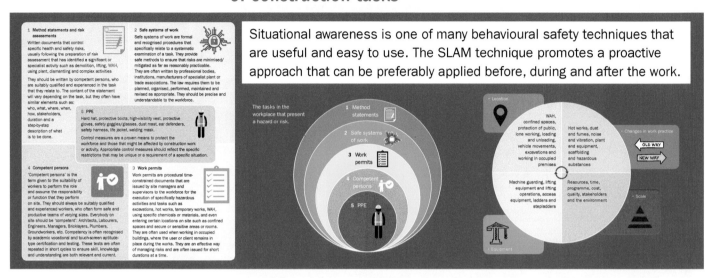

Situational awareness is one of many behavioural safety techniques that are useful and easy to use. The SLAM technique promotes a proactive approach that can be preferably applied before, during and after the work.

LINK

For more about the SLAM technique see pages 42 and 49.

LINK

For more about the cycle of compliance see pages 41–42.

'Community Projects', https://www.youtube.com/watch?v=QRXSEPZARIQ.

In previous units you have looked at how a cycle of compliance operating on site is a desirable, safe, effective and productive way of delivering a sustainable construction process. You should understand that construction managers, supervisors and the workforce work together using administrative procedures, safe systems of work and situational awareness techniques to mitigate accidents, incidents and near misses.

You have also considered how employers and employees work together with a common objective to deliver safe, productive construction in a secure and timely fashion through organised and professional procedures, environmental management of the workplace and within the culture of organisations/teams.

It is now reasonable to conclude that if the workforce applies positive health and safety practices in the completion of construction tasks, using good quality tools, techniques and materials, then a high-quality positive outcome is more likely.

However, this is only possible if the same stakeholders ensure that their direct, personal and continuous approach to the following three elements is delivered as a matter of routine. Only then can you be sure that there is a compliant and consistent cycle of full compliance while working, until such time as the task in completed:

1 Cleanliness and safety of work area
2 Safe working practices
3 Use of correct PPE.

LINK

Remember the iceberg poster on page 59?

SKILLS It is essential that you **can do a task well** because your own application of expertise is always **protected** when exposed to potential harm in the workplace. If you protect yourself with the correct use of the **appropriate PPE**, while working to the **approved methods** in a **clean (and often shared) environment** that you have effectively and reasonably taken adequate steps to protect yourself and each other from risk while working, then you will be a lot safer.

KNOWLEDGE / ## Cleanliness and safety of work area

A familiar behavioural safety sign seen on many construction sites all over the world often reads, in effect, 'a clean site is a safe site'. This aspirational sign is intended to promote the workforce and stakeholders to keep their activity and immediate work area clean from waste products and clutter. The cleanliness of the site is also often mentioned within the contract documentation and pre-construction information provided by the client. This is for reasons of best practice and maintaining high standards, and because the law states that construction sites and every workplace must be kept clean and in good order. Additionally, contractors must plan, manage and monitor to ensure work is carried out safely and without risks to health. So, this means that there should be a plan in place on how the site is cleaned, kept tidy and how the housekeeping is to be achieved. This is because waste is a risk unless it is managed and controlled.

KEY TERMS

Housekeeping: The approach to day-to-day operational cleanliness and storage.

Prefabrication: A manufacturing process that promotes the construction of building components prior to their installation on site.

Irrational storage: Storage that could cause an accident or or allow the materials being stored to be damaged.

APPLICATION / Examples of such waste and their potential to cause harm are shown in the following table. Copy the table into your notebook. Can you complete it?

Type of waste	Form	Potential harm	Control measure
Site dust	Layers of dust on floor such as sawdust and silica	Lung disease and varying levels of bronchial discomfort	Off-site cutting, prefabrication, dust extraction equipment at point of use, damping down and immediate reactive vacuum extraction
Off-cuts of timber and other materials left on the floor	Long and short lengths of materials and off-cuts left on walkways	Slips, trips and falls can disable and/or injure the workforce	Immediately pick-up or recover the waste materials and place them in a recycling skip or agreed point of disposal
Trailing electrical leads	Electrical leads left on the floor or at low level	Slips, trips and falls can disable and/or injure the workforce	Use cordless tools and/or ensure electrical leads are secured in cable trays or fixed above head height
Dirty and unkept toilet facilities			
Irrational storage of mixed materials not stored within a separate compound			

Dust extraction fitted to cordless drill.

Dust is transferred for controlled recycling or disposal.

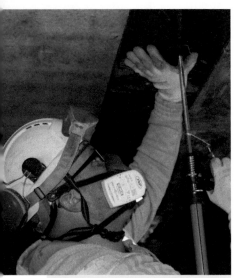

Always wear the correct PPE.

Waste is often put into a skip.

LINK

For more on legislation see
Unit 1.

KNOWLEDGE // # Prevention of accidents

In the UK, approximately 1,000 workers per year either fracture bones or suffer serious dislocation of joints due to tripping over while at work. Most of these incidents are preventable through managing materials and activities:

✓ Planned deliveries

✓ Designated storage areas

✓ Safe stacked storage at height (with adequate edge protection and access)

✓ Separate flammable material storage

✓ Organised pedestrian routes kept free of waste

✓ Tidiness in the workplace.

Waste management

Laws in relation to waste disposal are currently enforced by the Environment Agency and local authorities. At the earliest possible stage of a project, the contractor must decide:

* **How** wastes produced during the work will be managed in a timely and effective way.

* **Who** the responsible and licensed contractor is for collection and disposal of specific wastes produced on site. Early and decisive arrangements need to be in place at the earliest opportunity.

KNOWLEDGE // # Safe working practices

In Unit 1 the primary legislation that applies to the construction industry was discussed. With this knowledge in mind you can confidently enter the industry and potentially work in a safe environment if employers, employees and wider stakeholders practice the ethos that health and safety is everyone's responsibility.

You also learnt that change happens and no condition is permanent. The eight categories of work (listed below) identified to be relevant must now be respected by production of a site specification method statement and a task-specific risk assessment for approval prior to starting work.

1 Health and Safety at Work Act 1974

2 Reporting of Injuries, Diseases and Dangerous Occurrences Regulation 1995 (RIDDOR)

3 Control of Substances Hazardous to Health Regulation 2002 (COSHH)

4 Provision and Use of Work Equipment Regulations 1998 (PUWER)

5 Manual Handling Operations Regulations 1992

6 Personal Protective Equipment at Work Regulations 2002 (PPER)

7 Working at Height Regulations 2005

8 The Control of Asbestos Regulations 2012.

On starting work, you must continually monitor, assess and report abnormal activity and behaviours while working. The construction industry works at a fast pace and working environments change rapidly, so you must always remain alert and on guard.

If a situation doesn't look or feel quite right, **then it is likely not**. If you experience these situations or have feelings such as these, **then don't start working, or stop work and report it to a supervisor or manager**.

By continual application of practical and simple control measures when working on site, positive outcomes will be more likely. All eight categories of working should be covered within an approved risk assessment and method statement (RAMS). Generally, though, the course attendance and collaborative approach stakeholders work confidently and compliantly in a shared workplace, alongside the certification shown in the table below.

Category of work	Included with RAMS	Certification of basic understanding	Specific course attendance	Toolbox talk
Health and Safety at Work Act 1974	✓	✓		
Reporting of Injuries, Diseases and Dangerous Occurrences Regulation 1995 (RIDDOR)	✓	✓		✓
Control of Substances Hazardous to Health Regulation 2002 (COSHH)	✓	✓		✓
Provision and Use of Work Equipment Regulations 1998 (PUWER)	✓	✓		✓
Manual Handling Operations Regulations 1992	✓	✓	✓	✓
Personal Protective Equipment at Work Regulations 1992 (PPER)	✓	✓	✓	✓
Working at Height Regulations 2005	✓	✓	✓	✓
The Control of Asbestos Regulations 2012	✓	✓	✓	✓

KNOWLEDGE / Use of correct PPE

PPE's widespread use in the industry is often no longer task specific, as many organisations (particularly large businesses) apply strict procedures and rules demanding that protective hats, gloves, footwear and high visibility clothing are worn everywhere other than in the site accommodation. Such rules are put in place on the basis that they have a greater impact on accident prevention statistics.

However, some people suggest that each task should be considered separately and PPE chosen for the respective job. This makes workers think about what PPE they should be wearing and not take it for granted. Also, it is more sustainable, as fewer resources would be used. PPE should only be relied upon as a last resort and should be worn from the start of a job.

Personal protective equipment must be worn at all times

Effective application of PPE includes the provision (issued free by employers) of instructions, procedures, training and supervision to encourage people to work safely and responsibly.

Even where engineering controls and safe systems of work have been applied, hazards often remain in the form of residual risks.

Stakeholders remain at risk both in and adjacent to the workplace, so lungs, feet, heads, hands, eyes, skins and bodies all remain vulnerable unless protected. Exposure to the elements and extremes of temperature also remains a hazard.

Workers' hands can get very dirty.

Downloadable

The Illuminate Constructing the Built Environment dedicated website, https://www.illuminatepublishing.com/built_environment.

APPLICATION / Using the list below, identify what part of the body is affected by the related industrial disease/disorder contracted while working in the construction industry.

1 Asthma

2 Chronic bronchitis or emphysema – also known as chronic obstructive pulmonary disease (COPD)

3 Deafness

4 Dermatitis

5 Pneumoconiosis (including silicosis and asbestosis)

6 Osteoarthritis of the knee in coal miners

7 Prescribed disease A11 (previously known as vibration white finger)

8 Mesothelioma and other asbestos-related illnesses

9 Musculoskeletal disorders

10 Pierced cornea

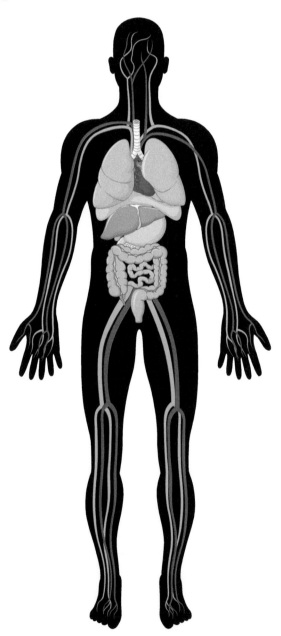

AC3.3 Evaluate quality of construction tasks

The construction industry has always had issues with negative outcomes. These issues often relate to untimely delivery, poor health and safety, image and poor quality. If the individuals within the industry improve their delivery standards, then these basic outcomes will also improve and so will those of the industry.

SKILLS / Evaluation

By introducing process and procedures into your daily working routine you can greatly improve the outcomes of health and safety, the programme, quality, cost and image. In doing so, the construction industry has an increased chance of greater sustainability.

KNOWLEDGE / Use the skills, knowledge and understanding that you have learnt and apply them to your projects.

UNDERSTANDING / When you consistently apply what you have learnt, learn what aspects can be improved upon and then apply these improvements so that you make the most of **your** three precious things:

LINK

For more on health, time and resources see page 31.

1 Health

2 Time

3 Resources

APPLICATION / Complete a process quality check sheet, as seen on the next two pages, for every task that you execute. They can be downloaded from the dedicated website. Learn from them and continually improve your skills.

The Illuminate Constructing the Built Environment dedicated website, https://www. illuminatepublishing.com/built_environment.

Site-specific Quality Check Sheet   Location   Date   /   /

	🗺 Location on plan (also refer to layout overleaf if not indicated here):	
1	⚡ RAMS in place?	Consider if they are an issue to this or other work ongoing. Check for danger while you work.
2	✛ Position/orientation on grid	Consider the relationship to your work in relation to setting out.
3	🔍 Line	Consider how the line of this element impacts of other proposed lines. Consider impact on cutting or modular dimensions of other materials.
4	📏 Level	Consider how the level of this element impacts on proposed adjacent levels. Consider impact on other adjacent work.
5	📄 Certification and testing	Consider if your work needs to be photographed or signed-off.
6	⚙ Execution and approach	Is the method agreed? Right skill set deployed? Sensitive area?
7	🗔 Construction detail compliance	Has the detail been followed or proposed alternative solutions agreed with team?
8	🧹 Cleanliness and protection	Can it be cleaned up now? Is cleanliness acceptable? Can it be better? Covered skips? Perimeter clean? Is protection required?
9	⚠ Public protection, work slips, trips and falls	Can the scope be reduced to make ramps less of an issue? Is the material to make-good ordered? Is there a requirement for other material on site, cleaning-up ready for emergencies?
10	🕐 Timeliness	Enough materials, resources and stakeholder buy-in present?
11	🧩 Next phases	Consider implications on next phases? Is there a better sequence available? Do you need to report anomalies (irregularities or inconsistencies) up-line?

List the best aspects of your work against the success criteria:

1

2

3

4

5

List three aspects of your work that could be improved:

1

2

3

Your conclusion:

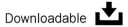
Compliance check sheet for tasks

- **Health and safety**: Are you and those adjacent to you **working safely** (your number one priority)?
- **Technical information**: **Accurately interpret** data and information from **more than one** source or type.
- Plan the **sequence** of work using logical timescales.
- Accurately **identify** and **specify** the resources required for the task.
- Accurately **calculate** and **record** all the materials required for the task (use standard conventions).
- **Brief**: Ensure that the success criteria are recorded on the previous page and the **best aspects** of your work correspond.
- Effectively complete your **preparatory tasks in a logical** sequence.
- Complete **your** task **independently**.
- Effectively use a **range** of **techniques** to complete your task.
- **Tolerances** of the work are compliant (**are you satisfied** with the **quality** of the completed task?).
- Final check that the quality of the work is compliant with:
  1 Success criteria (tolerances, timescales, selection, quality, what good looks like)
  2 Specification
  3 YOUR OWN JUDGEMENT.

Additional notes (if required)

_____

_____

_____

_____

_____

_____

_____

_____

Sketch (if required)

Planning Construction Projects

LO1 Know job roles involved in realising construction and built environment projects

AC1.1 Describe activities of those involved in construction projects

KNOWLEDGE / **Client's team (client, Architect, Engineer, Quantity Surveyor, project manager, designer)**

Client

KNOWLEDGE / The client is essentially the customer of the entire construction industry. This customer will be responsible for appointing a team of construction professionals who will help each other plan, design, construct and ultimately maintain the development, often a new building, extension or other form of works such as a refurbishment or asset. Often through this team the client will either delegate administration responsibilities or administer the contract themselves and ensure that the wider team is paid in a timely and consistent manner. The activity of the developer is often determined by the amount of investment they can raise in the form of savings, loans and tenancy agreements with potential individuals or organisations that wish to rent or buy part or all of the building or development.

UNDERSTANDING / Clients vary in size, which is reflected by their activity, for example:

- A family who owns a small domestic property may wish to refurbish or extend it to cope with a growing family or enhance its features to raise its market value.

- A person or family may wish to construct a new home themselves and embark on a 'self-build' or maybe employ a builder to construct it.

- A local authority may want to construct a new school to cope with a growing population in its area.

- A large commercial developer may want to redevelop a large but run-down or dated area within a city centre and utilise the space for mixed use developments such as studios, offices, shops and accommodation/private apartments.

- A larger organisation such as a pension trust fund or wealthy entrepreneur may construct a 'hyper-project' such as a data centre to sell or lease to a technology company to store data or act as an administrative base.

Architect

KNOWLEDGE The Architect will advise the client and team on the design (look and feel) of the building or development. They will agree to work within a certain timeframe (programme) to plan, design and often to check quality and the compliance systems of the rest of the design and construction team; when doing this role, they are often referred as the 'lead' designer. This designation of increased responsibility comes with a cost for the client, reflecting the increase in time and liability to the Architect and its own internal team. That said, some Architects will only be appointed by the client to deliver basic design information on a rolling basis in instances where planning permission and approval are uncertain.

More recently, Architects have been known by slightly different titles, reflecting the transition to the digital environment. Here BIM has been promoted by the government and larger developers such as the Ministry of Defence to help manage huge estates often spread across different regions and, in some instances, different countries. The focus on varying levels of BIM compliance can often lead to intense, and time-consuming, data-rich modelling that aids the planning, design, construction and particularly maintenance of the building and/or development.

UNDERSTANDING Architects are now sometimes called Architectural Technologists, Architectural Consultants or various similar role-names within that particular architectural practice. Despite all these titles, the Architect is generally the most senior person within the practice and ultimately responsible for delivery of the form and functional aspects of the design, and is capable of communicating with all other stakeholders in the industry. The Architect's activities will depend on the scope of their appointment with the client and the discharging of their associated responsibilities to deliver parts, or the whole of those defined in the RIBA plan of work. The activities often go well beyond the traditional responsibilites you might expect of this role.

Engineer

KNOWLEDGE The Engineer's activities can be very similar to those of the Architect. The primary activity focuses on the discharging of planning conditions generated by the planning process that often get engrossed into the project designer's scope of works and appointments. In such cases, the planning authority may place certain, defined improvements as part of the project that demand the Engineering expertise of structural and civil engineering professionals. These engineering roles are highly specialised. As the name implies, Structural Engineers focus on the structural design and stability of the building and development, ensuring that snow, wind and other environmental loadings are accommodated within the design.

UNDERSTANDING Civil Engineers focus on drainage systems, road construction, complex ground investigations and foundation solutions that will work in the best interests of the building and local environment. They work closely with their structural engineering colleagues to plan, design and check the quality of the construction process.

Quantity Surveyor (QS)

KNOWLEDGE Quantity Surveyors (QSs) are often referred to as cost consultants and sometimes commercial managers. Their activities include procurement and appointment of a competent builder or specialist contractor to deliver portions of the works such as bricklaying, plastering, carpentry and mechanical and electrical contractors.

UNDERSTANDING QSs remain active and vigilant during the project by coordinating with the construction team to ensure that the performance of those they have appointed provide a compliant, safe, timely and satisfactory quality of work. They ensure that the main contractor is paid by the client on time and that the supply chain (contractors and stakeholders who are appointed) are paid in a timely fashion.

Project manager

KNOWLEDGE The project manager (PM) is an advisor directly employed by the client. Their activity is focused around the monitoring and reporting of design compliance, commercial and operational readiness and capability of the contractor. They are responsible for the delivery of the project to the client in a safe and timely manner. This role is not to be confused with that of the contractor's PM (who is generally responsible for delivering the project on behalf of the contractor).

UNDERSTANDING The PM will focus on risks to delivery of the project objectives and commercial targets to ensure that a building, project, refurbishment or other investment that the client has procured stands the best chance of success. They give professional advice and make observations to collaboratively find the optimum solution to benefit the client.

PMs do not directly do any of the work in relation to the project themselves, so they operate primarily as observers who report regularly in the form of written reports to their employer (often the client) directly. PMs are authorised and therefore empowered by the client (on their behalf) to authorise payments, provide instructions and variations, and conclude when a project is suitable for handover.

PMs can add a tremendous amount of value to a project as it approaches completion, by monitoring progress and managing and coordinating the end-users of the building to ensure that handover and first use is a success. Their activity is governed and should be a healthy balance of acting on behalf of the client in a fair and reasonable manner while executing the contract collaboratively with the contractor and stakeholders.

Designer

KNOWLEDGE The term 'designer' can be applied to any member of the consultant team who has a function for the planning and realisation of a potential concept for the project. Traditionally, designers have been Architects, Engineers, and mechanical and electrical experts.

As the industry has evolved and become specialised, designers (who also have certain liabilities for their designs and concepts) operate in a varied and diverse supply chain of kitchen, window, door, roof, foundation, piling, interior and building envelope disciplines and many more special and technically advanced organisations.

UNDERSTANDING These designers ultimately represent a vision of the client's aspirations through their technical knowledge and grasp of the limitations of the products they promote, to ultimately help deliver compliant and sustainable building design solutions. More recently, designers produce data-rich modelling to represent their concepts in 3D and **4D** to remain competitive and comply with the government's desires for BIM conformity.

Designers' activities also focus on their obligations to safely deliver their concepts in context of the **CDM regulations**, ensuring that they do not compromise any part the building's construction or future function in terms of maintenance. Designers also have a professional responsibility to promote sustainable features that include recycled products, energy efficiency and thermal efficiency, and can themselves be recycled in the future.

KEY TERMS

4D: Four-dimensional; 4D information relates to data such as time (when and where details were formed or building components such as lighting, control panels or roof lights were installed) and programme (where and when proposed activities or series of linked activities such as steel work and its related floor construction is to be or has been formed). This allows the construction team to plan the construction process or the client and user to see details of the construction process after the building is completed. This can help them identify defects and resolve maintenance issues.

CDM regulations: Legislation that promotes the project team to:

- sensibly plan the work so the risks involved are managed from start to finish
- have the right people for the right job at the right time
- cooperate and coordinate your work with others
- have the right information about the risks and how they are being managed
- communicate this information effectively to those who need to know
- consult and engage with workers about the risks and how they are being managed.

(Source: http://www.hse.gov.uk/construction/cdm/2015/index.htm)

Contractor's team (builder/Site Engineer, site supervisor, safety officer, tradesperson, specialist sub-contractor)

Builder

KNOWLEDGE The contractor is represented on site by a singular person who can ensure the realisation of the concept in a safe, timely, efficient, productive and profitable approach ... the builder. This broad term defines a construction professional who can operate across the entire industry and more commonly and more recently is known as the construction, site or works manager.

This specialised professional will be activity delivering safe systems of work, and compliant and timely building work through an experienced and often pragmatic approach. Their activity is represented by a variety of tools and techniques that promote exemplary leadership, management, administrational and interpersonal skills as well as continual improvement.

> **KEY TERM**
>
> Exemplary: An example of the best of its kind.

UNDERSTANDING Builders often have to react quickly to rapidly changing scenarios and situations that are determined by fluctuations in the availability of labour, plant and materials, design changes, and instructions from the client, while working to constraints such as planning conditions and institutional standards (such as building regulations).

The builder should have a strong awareness of the work being done and good communication and technical skills, together with the knowledge, understanding and ability to be able to translate complex, technical data to others in a simple and understandable way. The term builder has increasingly been replaced by this type of builder to cope with the increasing complexity of the role.

Site Engineer

KNOWLEDGE The Engineer's activity is firmly centred around the responsibilities for the dimensional setting out and subsequent accuracy that follows. The Site Engineer may be activity engaged in several projects running concurrently or may be responsible for one larger project.

This activity will need the application of a highly skilled technician who should focus as much on safety as production to ensure a sustainable and positive construction period. Like the site manager, the Site Engineer needs to apply excellent communication skills relating to complex and critical information on the positioning of the building, its components and any of the associated drainage and infrastructure.

> **KEY TERM**
>
> Infrastructure: Important building and transportation networks.

UNDERSTANDING / The Site Engineer will be actively checking the data and recording 'as-build' information of any new and existing services that may be present. This data will then become part of the digital model and/or as-built information. The Site Engineer may also help coordinate the programme with the site manger to ensure the optimum production is achieved by stringent and proactive procedures involving working ahead of the supply chain (groundworkers, steel erectors and concrete specialists) to ensure accuracy, production, quality and timeliness.

Safety officer

KNOWLEDGE / The safety officer is an occupational health and safety professional who regularly visits the site. The activities of this person focus on the compliance of the project with health and safety law and how it is being realised on site, generally through the processes and procedures of the organisation. Their activity is usually one of 'observe and report', although more recently they have become proactive.

UNDERSTANDING / Part of this new proactive approach has been to record and utilise the data that they collect to prevent accidents, incidents and near misses from occurring in the future. They also help execute behavioural change workshops, safety campaigns and toolbox talks with the site-based team.

They conclude their site inspections in factual reports that are distributed to the site team and senior managers within the organisation.

Tradesperson

KNOWLEDGE / The crucial and valuable skills, knowledge, understanding and application of a diverse range of trades are vital to the construction industry. Tradespeople's current activities are focused on the provision of the high-quality and cost-effective application of trades such as groundworks, bricklaying, steel erection, roofing, plastering/dry-lining, decoration and highly specialised arts and crafts such as thatching and drystone walling to name a few.

UNDERSTANDING / These tradespeople engage in sub-contract agreements and sometimes work directly for larger building contractors. Their activity is nationwide, so they may travel long distances to ensure sustainability. Activity of the tradespeople can often be via direct relationships with clients who may even compel the main contractor to utilise them. In instances where there is a failure such as a crack, leak or defect, a tradesperson often only guarantees their own work as opposed to any associated failure of the materials.

Specialist sub-contractors

KNOWLEDGE // Like the tradespeople, specialist sub-contractors are often highly skilled and technically advanced contractors who are employed by the main contractor or client to offer design, manufacture, and construct and sometimes maintain a specialist component or element of the building or development. These elements or components are often:

- piling/foundation solutions
- structural steel and concrete frames
- roofing membranes
- windows and curtain walling
- coatings and waterproof membranes
- finishes and furniture.

UNDERSTANDING // Specialist sub-contractors' activities may often relate to the manufacture, supply and installation of these components to ensure the quality and viability of their products or services. This group has greater exposure to product failure and any direct implications that may result, such as structural cracking or roof/window leaks. When this occurs, they will react accordingly under the terms and conditions of any warranties that may be underwritten (such as an insurance policy to protect the product or installation). This warranty process can often be the main difference between a tradesperson and a specialist sub-contractor.

Statutory personnel (building control inspector, town planner, public health inspector)

Building control inspector

KNOWLEDGE The title of the building control inspector is also more commonly known as the building control officer (BCO), indicating the high level of statutory governance this position holds. The activity of a BCO is not only restricted to the checking function as traditionally conceptualised but also involves an exceptional understanding of how the current building regulations can be and should be applied in different scenarios.

The BCO can be employed directly by the local authority (public sector) or more recently by private accredited organisations (private sector). Both types of organisations are appointed by the contractor to help approve, in principle, drawings and specifications and also comment on the design well ahead of the construction phase (often during the feasibility stage). This degree of collaboration and early engagement helps the client, design and construction teams to adequately plan the work thoroughly in advance.

During the construction phase the BCO will visit the site under an agreed programme, with the contractor, to inspect certain elements of the works such as the formation level, gas membranes, damp-proof course level and so on, until such time as the BCO is convinced that the work is in accordance with the building regulations. It is not the BCO's responsibility to check specifications and products, as these can be varied and subjective.

UNDERSTANDING The activity of the BCO is both diverse and multi-disciplinary, as they must have a broad knowledge and understanding of everything from structure to mechanical and electrical services and how they perform in the context of the building when there is an emergency such as a fire or natural disaster. The building regulations exist primarily to protect the lives of the occupants, users and public. BCOs comment via a report on their findings during regular inspections, when they critique the formed construction details such as weathering, thermal efficiency, potential performance in fire and provision for those who use wheelchairs or are visually impaired.

They are highly respected construction professionals who have the last word on matters of compliance and quality.

Town planner

KNOWLEDGE Town planners work actively across the UK, generally employed by local authorities but can also work privately as planning consultants. Town planners are also referred to by other titles such as regional and city planners and urban technologists. Their activity is primarily strategising on behalf of local government to ensure that both local land resources and central government rules, regulations and law are complied with on matters of development and redevelopment. These construction-related professionals are important in improving the built environment.

The town planner ensures that appropriate construction, refurbishment and infrastructure projects fulfil all the criteria of national and local guidance in readiness for their subsequent processing through the planning system. During this period, town planners may impose certain conditions that have been well considered and so are beneficial to existing and new communities. For example, a planning application for a large-scale housing project may be under consideration from a housing developer. The homes are needed urgently for a young and fast-growing population, so the town planner may impose a planning condition on the development that requires a new primary school to be included as part of the proposed development at the expense of the developer.

KEY TERM

Validate: Prove the accuracy.

Discharge: The successful process of ensuring an obligation such as a planning condition is achieved.

UNDERSTANDING The decisions made at this stage are crucial to the futures of the communities, the built environment, and the country's wellbeing and prosperity.

Town planners also spend large portions of their time processing appeals, new applications, giving advice and collaborating with other agencies, so communities are prepared for the future (see mind map below).

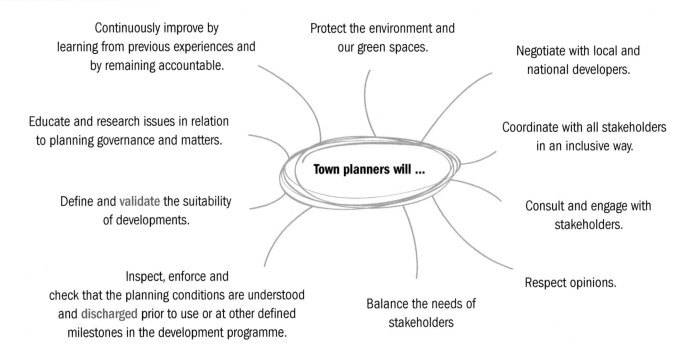

Continuously improve by learning from previous experiences and by remaining accountable.

Protect the environment and our green spaces.

Negotiate with local and national developers.

Educate and research issues in relation to planning governance and matters.

Coordinate with all stakeholders in an inclusive way.

Town planners will ...

Define and validate the suitability of developments.

Consult and engage with stakeholders.

Inspect, enforce and check that the planning conditions are understood and discharged prior to use or at other defined milestones in the development programme.

Balance the needs of stakeholders

Respect opinions.

Public health inspector

KNOWLEDGE // These are independent professionals often employed by local authorities and central government to ensure that local and national environmental standards of living, working, resting and leisure are achieved and maintained. These professionals operate across all sectors and focus on compliance by local authorities, businesses and residents to maintain health, safety and environmental rules, regulations and laws.

UNDERSTANDING // These professionals actively police the impact on the environment by focusing on potential sources of chemical and biological contamination, noise, air and water pollution, and when there is grievance from a third party made to the local authorities (see mind map below). Their activity is focused on the protection and wellbeing of those that share local, national and sometimes global communities with the intention of continuously improving society's approach and experiences on all health and wellbeing matters.

Test and inspect vigorously, using both traditional and latest science techniques and technologies.

Continuously improve by learning from previous experiences while remaining accountable.

Public health inspectors will ...

Consult and engage with stakeholders.

Educate and research issues in relation to environmental and health governance law.

Respect opinions and cultures of stakeholders.

This diverse and respected group of health surveillance professionals, comprising of doctors, nurses, executives, technicians and construction professionals, protect our immediate and long-term futures, ensuring that the built environment is 'fit for purpose' and prepared for the worst.

General (administrator, finance officer, public liaison officer, purchasing/procurement officer, caterer, security) construction projects and refurbishments and extensions

Administrator

KNOWLEDGE An administrator operates from within and across the many different organisations associated with the construction industry. They are often managers who ensure that the standards of the relative organisation's procedures and processes are compliant to the needs of that business or activity.

Administrators collect outputs and compliance in the form of letters, completed forms, emails and correspondence, and ensure that the responsible people complete the related administrative tasks or provide any outstanding information.

UNDERSTANDING The administrator then provides directors, senior managers and department leaders with this information, allowing them to understand any current issues or areas of improvement required to maintain the day-to-day running of the business. This provision includes the distribution of memos from directors to the workforce, pending important visits and monthly returns of operational or commercial significance.

Finance officer

KNOWLEDGE The finance officer is a senior figure within a regional business or organisation. They actively collect commercial outputs such as profit margins and turnover produced by the individual construction projects in readiness to process them into a monthly accounts report.

The finance officer also analyses the forthcoming periods of proposed project buying activity to ensure that the business's head office, represented by the finance director, is prepared to provide the adequate payments to allow the regional supply chain to be paid on time.

The finance officer works closely with the site/project managers and Quantity Surveyors to help them target areas of commercial improvement and transparency to ensure there are enough funds available at the right time to avoid litigation and maximise operational efficiency.

UNDERSTANDING Finance officers often work to a 12-month financial year, which often commences in April. The financial year is subdivided into quarters, as this helps the wider business to focus on achieving commercial goals and the commercial directors' plan.

Public liaison officer (PLO)

KNOWLEDGE The liaison officer is the public face of the organisation who is well trained in communications techniques. These techniques are intended to proactively inform the general public, clients and wider stakeholders of the development and any social or environmental impacts of it on a day-to-day basis. The liaison officer is also prepared to answer queries from the press or media and support senior managers when they represent the organisation on television, radio or when writing articles for the press. In a rapidly changing and multimedia covered industry, the PLO is likely to use social media channels to help promote the organisation or monitor and report media trends that may have an effect on the reputation of the specific project or business.

The purchasing/ procurement officer

KNOWLEDGE During the pre-construction period, the purchasing and procurement officer will advertise all opportunities for tradespeople, designers and specialist contractors who may offer goods or services that could benefit them and the project. During this time the position is often associated with local government employees who work within national and European procurement and government guidelines, to ensure fairness and transparency.

During the construction period the client or contractor will appoint suitable supply chain members who have been successful in satisfying the tender process. In larger organisations this role is often performed by a buyer but in most cases the commercial manager or Quantity Surveyor supported by the site team will perform the role.

Caterer

KNOWLEDGE The caterer is a vital hospitality professional who provides refreshments, snacks and lunches for the workforce, both in the office and on site when a larger construction project is commissioned. The caterer often has their own business and employees, all of whom will have food preparation qualifications.

The caterer often responds to orders at short notice and provides services for client progress/design meetings, busy construction sites and business development events.

The caterer's activity is often focused on providing the best value food, refreshments and services in a very competitive area of expertise.

Security

KNOWLEDGE As discussed in the previous units, the risk to the construction industry of fire, theft and vandalism is never far away. The construction security industry specialises in the asset protection of the site, offices, vehicles and workforce. Security services monitor the business for exposure to all these risks through the provision, use and maintenance of state-of-the-art electronic surveillance systems and a strong physical presence on site.

More recently, some specialist security organisations offer the service of drug and alcohol testing of the workforce, which is a rapidly growing national issue.

UNDERSTANDING Use of recreational drugs and alcohol by the workforce presents a hazard to the whole industry, as it greatly impacts an individual's ability and performance while at work. Such abuse reduces the workers' powers of situational awareness and therefore elevates the risk to themselves, colleagues and wider stakeholders.

KNOWLEDGE ## Refurbishments and extensions

In principle, the activities of these roles within the construction industry remain the same or very similar regardless of whether the project is a refurbishment, extension, new build or infrastructure project. If the organisation has contracted to deliver this project then, regardless of the size or scale of the project, the roles and obligations often remain relevant to the task at hand.

AC1.2 Describe responsibilities of those involved in construction projects

KEY TERMS

Defective: Work or a system that does not work as it was designed to do or has relevant flaws within it and is therefore non-compliant.

Statutory application: The formal process of submitting an official procedural form or adhering to a recognised process.

KNOWLEDGE and **UNDERSTANDING** Client's team (client, Architect, Engineer, Quantity Surveyor, project manager, designer)

Client responsibilities

- Fulfil their duties under the CDM regulations.
- Satisfy themselves that the project is feasible in commercial, practical and beneficial terms.
- Choose and appoint suitably qualified and experienced construction professionals by following the advice of their project manager.
- Trust the selected team to procure, design, construct and ultimately deliver the appropriate solution.
- Regularly pre-determine what the possible risks are that may threaten the development and act when appropriate.

Architect responsibilities

The responsibilities of an Architect who is appointed by a client are agreed prior to commencing any form of work by the Architect, to ensure that both stakeholders fully understand and accept who is responsible for delivery of what can be a long and expensive process, as shown in the following mind map:

Advise the client on any negative aspects of the construction process, such as defective work, risks, likely programme issues or delays.

Fulfil their duties under the CDM regulations.

Advise the client on the feasibility, design and procurement processes and costs associated with the proposed development.

Apply for planning permissions and building regulations approvals.

An Architect will ...

Advise the client on completion and handover and practical future maintenance routines, and provide final drawings and records.

Advise the client on design solutions in the form of concept and detailed designs that conform to current institutional standards, and in particular the current building regulations.

Administer the contract for the client by chairing meetings, issuing formal communications and instructions to team members and setting high standards for them to reach.

Advise the client on the selection of the other suitable team members; once the team is assembled the Architect will then coordinate and collaborate with them (ideally, as early in the process as practical).

Prepare any statutory application for elements such as water, gas, electricity and data provision.

Engineer responsibilities

KNOWLEDGE / and **UNDERSTANDING** /

- Fulfil their duties under the CDM regulations.
- Advise the client team on the feasibility, design and procurement processes and costs associated with the proposed development in terms of geotechnical, ground investigations and best value structural solutions.
- Advise on the engineering risks to the project.
- Provide the Architect with an analysis (full details) for planning permissions and building regulations approvals.
- Advise the client team on design solutions in the form of concept and detailed designs that conform to current institutional standards, and in particular the current building regulations.
- Advise the client team on the selection of the other suitable methods, products and specialists that may be needed to provide specialist engineering solutions to challenging projects, such as poor load-bearing ground, reclaimed ground that is contaminated or constraints where there are existing buildings or infrastructure.

Advise the client on completion and handover about practical future maintenance routines, and provide final drawings and records.

Prepare any statutory application for elements such as foul sewers and storm-water applications and easements where there are existing services.

Engineers ...

Advise the client on any negative aspects of the construction process such as defective work, risks, likely programme issues or delays.

Advise the Architect and client during the contract by attending meetings, issuing formal communications and advice to the team members and setting high standards during the checking and test/inspection period.

KEY TERMS

Foul sewers: The network of foul (human waste) drainage that exists in the UK to transport human waste to treatment works where the waste is removed and the water treated.

Easements: The legal right to cross over, near or by another party's land or part of their infrastructure, like a gas or water main, for a reason.

Quantity Surveyor (QS) responsibilities

KNOWLEDGE // and UNDERSTANDING /

- Appoint the contractor formally by administering all the contractual documentation in readiness for the client project manager to check and discharge formally through the contract terms and conditions.
- Coordinate and administer any special warranties that are required to protect the client from defective design, materials/products and workmanship in readiness for the project manager to check and approve.
- Provide an up-to-date cash flow forecast for the client so they are aware of the value and frequency of payments during the project's lifecycle.
- Attend project meetings to collaborate with the designers and operations team so they are aware as early as practically possible of any potential commercial risks. Then act to mitigate these risks, ensuring the client is always aware of them and the potential commercial cost, such as additional works instructed by the design ream, shortages in resources and any other factors that could increase costs or time to the project.
- Produce a regular (usually monthly) detailed cost report that summarises the commercial activity of the previous period and forecasts the next period's anticipated workflow.
- Ensure the contractors are paid on time and as per the conditions in the contract but only after the QS has checked that the work has been completed, checked the work has been done correctly and signed it off.

Project manager (PM) responsibilities

KNOWLEDGE // and UNDERSTANDING /

- Represent the client during the entire plan of work and ensure the client is aware of their duties under the CDM regulations.
- Ensure that the client's interests are protected during the entire project lifecycle.
- Ensure that the project is delivered on time.
- Ensure that the project is delivered on budget.
- Ensure that the project is delivered by competent persons in a safe manner.
- Make sure the client is fully aware of all commercial and operational risks and potential opportunities on a regular basis.
- Manage and deliver the handover and occupation of the project by collaborating with the contractor and end-building users.
- Ensure that the contractor has planned adequate resources to deliver the project.
- Produce and monitor key performance indicators (KPIs) during the construction phase that indicate how the team are performing, including points such as the number of complaints from local populations, number of accidents on site, communication skills of the team and any other agreed deliverables that the client considered important.
- Issue timely instructions, responses to requests for information and responses to queries. Change controls and attend/record all meetings on behalf of the client.

Designer responsibilities

KNOWLEDGE and UNDERSTANDING

- Plan, manage, monitor and coordinate health and safety from the time they are appointed and join the design team by identifying, eliminating or controlling all foreseeable risks. For example, by not specifying needlessly heavy or large products or maybe by suggesting that products could be assembled off site by the manufacturer. Some designers even reduce on-site cutting by specifying pre-cut or modular materials.

- Prepare and provide relevant information to other dutyholders in a timely fashion such as 3D models that the designer has checked for clashes with other building components prior to issuing and before they are installed. This includes understanding the importance of providing design information on time to the contractor so they can procure and install a component on time. For example, some mechanical and electrical products have extremely long lead times.

UNDERSTANDING

- Provide all relevant information to the principal contractor to help them plan, manage, monitor and coordinate health and safety in the construction phase by ensuring that what they design can be feasible to form by not designing complex details that rely on working at height or within confined spaces for long periods of time.

- Regularly attend meetings and make site inspections as the work is formed on site and comment on the quality of their findings prior to anything wrong being covered-up. For example, a complex mechanical and electrical installation above a suspended ceiling needs to be tested and inspected prior to the ceiling being constructed; or sections of buried drainage need inspection and testing prior to backfilling to ensure that they do not have any leaks as reworking deep drainage could be extremely costly.

Contractor's team (Site Engineer, site supervisor, safety officer, tradesperson, specialist sub-contractor), builder (site manager/construction manager/main contractors)

Site Engineer/builder responsibilities

KNOWLEDGE

* Translate, manage and organise setting-out criteria (often in the form of physical markings and reference points on site) between the sub-contractors, by supervising their ability to adhere to the dimensional tolerances expected from them. Report any deviations from these immediately to the site manager.

* Observe and report (to the site manager) all matters that relate to health and safety, quality, programme, progress and approaches to working that deviate from the sub-contractors' RAMSs or the construction phase plan.

* Ensure that any existing live services are plotted and signposted on site to avoid the sub-contractors hitting them ('cable strikes' are dangerous occurrences that can be avoided).

* Remain the accountable construction professional on site to define, monitor and report on all engineering-related tolerances and deviations.

UNDERSTANDING

* Collaborate with and encourage the workforce to deliver exemplary quality through the organisation's processes and procedures.

Site supervisor responsibilities

KNOWLEDGE

* Effectively translate and communicate how the workforce should work and what is expected from them while they work. The site induction is an opportunity to do this, as supervisors can be authorised to give these. Once the work has started, they can also give 'toolbox talks' to the workforce.

* Lead the team doing the task, ensuring they do it safely, productively and in a timely way.

* Remain the front-line accountable person who can immediately react to challenges and help the workforce to stay safe, productive and engaged in the construction process.

UNDERSTANDING

* Report directly to the site manager on aspects and tasks that have been given to them. These tasks can range from supervising two or three operatives to clean, a newly completed area of a building or taking receipt of an important delivery that must be secured from loss or theft.

Safety officer responsibilities

KNOWLEDGE / and UNDERSTANDING /

- Observe and report to senior management and directors of the organisation all aspects and details of their site audits to enable the organisation, management and workforce to:
  - Help them identify workplace risks and effectively know how to deal with them.
  - Ensure that the health and safety controls measures are in place and appropriate to the task or situation being undertaken.
  - Work closely with everyone to show the level of commitment of the organisation and workforce to deliver a safe and compliant project environment.

Tradesperson responsibilities

KNOWLEDGE / and UNDERSTANDING /

- Comply with the directions and instructions given to them by the main contractor or the employer. These directions are often the same across many different sites and similar between other main contractors.
- Coordinate with other contractors or tradespeople by planning and managing their work to optimise health, safety, quality production and programme aspects of the works. It is often the transition between phases of work or different trades where issues arise, such as sharing areas of the works at the same time or potentially working over each other.
- Remain interested and helpful when consulted about matters that affect their health, safety and welfare by taking care of their own health and safety and others who may be affected by their actions or work. This includes reporting anything they see which is likely to be dangerous to their own or others' health and safety. Tradespeople should always remain cooperative with their employer, fellow workers, contractors and other dutyholders; for example, where different trades are sharing a canteen or office, they should remain respectful of each other and not antagonise co-workers, as this is detrimental to morale and could cause issues in the future.
- Strive to achieve the highest standards of quality that are relative to their craft, skill or trade. Once again, KPIs are an effective way to recognise exemplary quality together with trade awards and recognition.

Specialist sub-contractor responsibilities

KNOWLEDGE // and UNDERSTANDING /

- Comply with the directions and instructions given to them by the main contractor or the employer. These directions are often referred to in the contract that forms the basis of their relationship.

- Coordinate with other sub-contractors or tradespeople by planning and managing their work to optimise health and safety, quality production and programme aspects of the works. It is often the transition (move) between phases of work or different trades where issues arise, such as sharing areas of the works at the same time or potentially working over each other.

- Remain interested and helpful when consulted about matters that affect their health, safety and welfare by taking care of their own health and safety and others who may be affected by their actions or work. This includes reporting anything they see which is likely to endanger either their own or others' health and safety. All contractors should always remain cooperative with the main contractor, fellow workers, other sub-contractors and other dutyholders, for example where different trades are working on the same element such as a concrete frame they should agree a pattern of work that is safe and productive by not working above or under each other in case tools or materials are accidentally dropped, which could cause an accident or damage work.

- Strive to achieve the highest standards of quality that are relative to their craft, skill or trade. Once again, KPIs are an effective way to recognise exemplary quality together with trade awards and recognition.

- Attempt to refine their methods to improve safety and production such as prefabrication of components or mechanically cutting materials off site in greater controlled environments.

133

Site manager/construction manager/main contractors responsibilities

KNOWLEDGE

- Plan, manage, monitor and coordinate health and safety and production before and during the construction phase. The site manager will focus on liaising with the client, principal designer and other team members. The organisation's processes and procedures are used to do this and KPIs are also a great way to monitor performance.
- Ensure that the construction phase plan and quality plan are followed and regularly reviewed.
- Manage and organise cooperation between the sub-contractors by coordinating their work using tools such as programmers, leadership techniques and workshops.
- Deliver effective and engaging site inductions to the workforce and visitors. Some contractors use interactive presentations or video presentations to do this.
- Keep the site secure from unauthorised access and the workforce protected from theft of their possessions and data.
- Consult and engage the workforce to protect the health and safety standards on site. Meetings, workshops and feedback cards are an effective way to have meaningful consultation with the workforce.

UNDERSTANDING

- Look after the welfare of the workforce the provision of by ensuring the provision of clean, organised and has dedicated places to heat food, dry clothing, charge tools/devices and secure valuables.
- Maintain a diary and remind the team of why drawings and schedules are so vital to keep the project 'on programme' and pursue any missing or outstanding information.
- Keep accurate records relating to levels of labour, plant, materials and issues relating to the performance of sub-contractors. Communicate these issues and the implications of any negative aspects to the contractors.
- Remain the accountable construction professional on site that represents the contractor.

Statutory personnel (building inspector, town planner, public health inspector)

Building inspector responsibilities

- Give advice to the client on the current approved documents/building regulations. This can often be a set of marked-up drawings and comments on areas of improvement or where clarification of details or products is required by the Architect.
- Produce guidelines in relation to costs for submissions and how a building control submission is to be made. This will help the client to plan the construction process and allocate the costs.
- Scrutinise (check closely and carefully) and comment on applications and proposals to validate compliance with building regulations. This is similar to the first bullet point in this list but in more detail and as the works progress.
- Produce administrational reports, certificates and reviews for their employer and customers. This is to ensure that the team understand what any issues are and that they will react accordingly to resolve them.
- Regularly conduct site inspections of the works during key stages and report their findings/observations to the builder/client. This is crucial to fully understanding and appreciating the project in the context of the site.
- Produce and issue a final certificate that relates to the project. The client and contractor will need a formal certificate to agree that the works are completed. This is an important document as it defines closure of a legislative process. This can be vital to allow occupation and use of developments such as schools, hospitals and apartments, where many people share use of a building. This is because the risks associated with the threat of fire and safe access and egress (exit) are often easily compromised in an emergency.

Town planner responsibilities

- Help in the development of local plans and strategies of the local authority to best serve the people, environment and businesses of that area. This is done to ensure that the long-term viability and sustainability of these regions are not compromised. For example, areas of outstanding natural beauty should be protected but may need some modernisation to cope with increased visitors and population; while former industrial brownfield sites need to be regenerated by being developed but not overpopulated or have no provision of green spaces for the wellbeing of the population and users.
- Collaborate and work with central government, developers, local population and wider stakeholders to ensure the Town and Country Planning legislation is respected and upheld by processing planning applications fairly and inclusively.
- Remain accountable and balanced in their approaches to all planning applications.

KEY TERM

Brownfield site: A site that has previously been developed.

Public health inspector responsibilities

KNOWLEDGE / and UNDERSTANDING /

- Enforce the law in relation to matters concerning public health and health and safety in a variety of scenarios including all public, private, workplace and leisure-type environments. The inspectors do this by using their powers to:
  - Prosecute a person, business or other body who breaks health and safety laws.
  - Issue formal notices when improvements need to be made, such as where there is a dirty kitchen or when they discover staff who are not qualified to prepare food.
  - Interview people who may be able to provide evidence or information in relation to an incident or investigation.
  - Survey and inspect premises without prior notice (sometimes with a police escort). The inspector needs to remain safe, so may need protection or help during difficult scenarios.
  - Educate, support and offer guidance on how to comply with the law.
  - Remain the approachable and accountable professional, often employed by the local authorities to protect the interests of the public and the built environment.

Construction-related general support staff (administrator, finance officer, public liaison officer, purchasing/procurement officer, caterers, security) construction projects: refurbishments and extensions

Administrator responsibilities

KNOWLEDGE / and UNDERSTANDING /

- Ensure that the day-to-day communications within the organisation are operating effectively.
- Ensure that any reports or spreadsheets are stored/filed/distributed to the correct individuals and at the correct time.
- Remain approachable and collaborate with colleagues.

Finance officer responsibilities

KNOWLEDGE // and **UNDERSTANDING** //

- Ensure that the accurate and transparent current commercial positions of all projects are accurately represented in the monthly commercial report presented to the organisation's board of directors. This will allow the board of directors to act and formulate plans in the best interests of the organisation.
- Ensure that the monthly anticipated and projected cash flow is accurately forecast ahead of each quarter so there are enough funds to pay to the supply chain (sub-contractors and tradespeople).
- Remain the accountable person to lead, direct and coach the commercial managers (QSs), allowing them to remain commercially astute (be aware of the situation and use this to information to an advantage) and the organisation to profit at sustainable levels.

Public liaison officer responsibilities

KNOWLEDGE // and **UNDERSTANDING** //

- Ensure that the organisation is proactively public facing and accountable in the press, social media and to stakeholders and the public.
- Ensure that any press, social media or formal statement made by or on behalf of the organisation is approved by at least two board directors prior to release.
- Ensure that the organisation projects (shows to others) a consistent and positive image in terms of health, safety, environment, quality and educational/training matters.

Purchasing/procurement officer responsibilities

KNOWLEDGE // and **UNDERSTANDING** //

- Ensure that the local supply chain (within 200 miles by road) is made aware of all opportunities in relation to providing goods or services to all sites and projects that the business is responsible for.
- Ensure that the rules of the UK and European procurement law are followed when procuring goods or services on behalf of the organisation.
- Remain the accountable person for all matters relating to purchasing and procurement by keeping records to demonstrate that at least three quotations (summaries of the commercial value of the package from different sources or suppliers) have been received and fairly processed. This allows the team to compare the value for money of those that have supplied the quotations and the potential of engaging with them.
- Work closely with the public liaison officer to advertise and promote the opportunities available working with the business and remain the accountable person for all matters relating to purchasing and public procurement.

Caterer responsibilities

KNOWLEDGE // and UNDERSTANDING //

- Provide the highest standards of hygiene and food safety while servicing the organisation and its guests, visitors or other users.
- Ensure that the food offered is balanced, nutritious and inclusive to the needs of a diverse workforce. This includes serving food for people with food allergies and showing the food content for all meals that are served at the venues managed by the organisation.
- Ensure that all catering staff are adequately trained and qualified in the skills of food handling and preparation to both local authority and central government standards.

Security responsibilities

KNOWLEDGE // and UNDERSTANDING //

- Ensure that all the construction project sites have a secure perimeter to prevent members of the public and those who could potentially cause theft, vandalism or harm to the workforce are prevented from entering the site works and compound.
- ON ALL PROJECTS VALUED AT MORE THAN £5m CONTRACT SUM: provide a 24-hour, 365-day recorded closed-circuit television (CCTV) surveillance system on all sites and report all incidents or near misses within one hour of becoming aware of the incident (directly by telephone and email to the relative site manager) and a full-time security guard for ten weeks leading up to practical completion.
- ON ALL PROJECTS VALUED AT LESS THAN £5m CONTRACT SUM: provide an 'outside of normal working hours' electronic alarmed activation system with an attendant reactive visiting guard who can reach site within 20 minutes of activation for the duration of the work and a full-time, on-site security guard, for ten weeks leading up to practical completion.
- Install, monitor and maintain a biometric fingerprint reader that records all staff and worker movements on all projects regardless of value.
- Remain the accountable and proactive construction-related professional team that protects the interests of the organisation at all times by helping to prevent and monitor security- and safety-related incidents.

AC1.3 Describe outputs of those involved in realising construction projects

You have looked at the job roles of the individuals in detail and how these construction and construction-related professionals work alone and also together when they collaborate with one another. You need to understand the outputs of the various teams, which, after all, are the primary objectives of the project when attempting to realise any project or development.

KNOWLEDGE / and UNDERSTANDING / Client's team, contractor's team and statutory personnel

	Client's team	Contractor's team
Who?	Client, Architect, Engineer, Quantity Surveyor, project manager, designer.	Builder/Site Engineer, site supervisor, safety officer, tradespeople, specialist sub-contractors, site manager, construction manager, main contractors.
Overview	The client's team of professional designers and mangers will deliver the project by collaboratively working together.	The contractor's team of professional builders, Engineers, mangers and supply chain will deliver the project by collaboratively working with the client's team and statutory personnel.
Responsibilities	• Delivery of a project that is safely designed, procured and constructed under the legal duties such as current (2015) CDM regulations. • Appoint a competent contractor to help develop, realise and deliver the building and any associated works such as roads, landscaping and other planning considerations such as drainage and utility provision. • Deliver a project that is compliant to the planning conditions and building regulations relevant to the applications that were made in the early part of the project lifecycle. • Deliver a sustainable development that can be both useful and attractive to the local environment and somewhere where people and businesses can thrive.	• Delivery of a project that is constructed safely, as their legal responsibility states, to comply with legislation such as the CDM regulations, the 1974 Health and Safety at Work Act and the Construction Regulations. • Appoint and manage tradespeople and specialist sub-contractors to deliver a compliant high-quality building and related infrastructure project. • Deliver a project that is compliant to current and relevant institutional standards such as the building regulations and any approved codes of practice, while complying with any planning conditions such as restricted working hours and access constraints. • Deliver a project that is on time and on budget to the developer. • Maintain a record and develop lessons learnt procedures to ensure improvement on all matters relating to health, safety, environment and quality.

Statutory personnel	General construction projects
Building inspector, town planner, public health inspector.	Administrator, finance officer, public liaison officer, purchasing procurement officer, caterer, security.

The public interest is represented by the provision of a local authority professional team of inspectors, planner and health professionals who provide advice and approval of the current approved documents/building regulation, planning law, health regulations and essential institutional standards that affect and protect the public health, safety and wellbeing.

Building inspector responsibilities:
- Remain impartial on all matters relating to approval of plans and designs from developers and the public by being transparent and providing clear feedback by using meetings, letters, drawings and emails to communicate issues.
- Continue to advise and educate the public and developers on all matters relating to development and building control by producing clear information and guidelines about the relative legislation and processes.
- Produce detailed and timely reports in relation to decisions and requirements, ensuring they are communicated in a timely fashion to developers, the public and any stakeholders.
- Always consult the public and give them plenty of time and opportunity to review planning applications and any construction work that may affect them.

This broad and mutli-disciplined team of administrative staff, accounting, public relation, business development, procurement, catering and security professionals provide everything that the industry needs to operate efficiently and sustainably. They all help to plan, manage, support, check, maintain and protect the industry from harm and inefficiencies. This group is vital to maintaining effective and uninterrupted supply chains.

Administrator, finance officer, public liaison officer, purchasing/procurement officer:
- Produce timely and accurate commercial and operational reports that help the operational staff, commercial teams and company directors to deliver safe, timely and high-quality building stock while optimising profits.
- Fairly and efficiently communicate all potential current and future projects to local and national tradespeople, sub-contractors and supply chain members, so they have a fair chance to tender for all opportunities in working with the business.
- Ensure that sub-contracts are awarded not just on price alone but also by understanding what safety and quality value can be gained by working with tradespeople and other members of the supply chain.

Caterer and security:
- Help to provide the business with safe and sustainable services that protect the workers and business from theft and risks to their health, safety and security.
- Help to provide enjoyable, diverse and nutritious food and drink services to the highest standards possible to help the workforce work productively and improve their wellbeing.
- Strive to improve the quality and experiences of the business and stakeholders by learning from past experiences with the intention of improved levels of service provision.

LO2 Understand how built environment development projects are realised

AC2.1 Describe processes used in the built environment development projects

KEY TERM

106 agreements: Conditions imposed by the local authority planning department such as: for every 300 houses constructed the developer must provide a nursery school; or contribute to an adjacent road widening project; or maybe provide a pedestrian bridge over a local busy highway that the occupants of the new housing are likely to use. This ensures that any new development is safe and sustainable for people to live, work or play in and not subsidised by the tax payer.

Planning (design, project planning, procurement)

KNOWLEDGE and **UNDERSTANDING** Design

The design process is initially driven, managed and controlled by the design team, generally consisting of the Architect, Engineer, mechanical and electrical designers, together with other specialist consultants who help inform the design process such as acousticians, geotechnical experts and environmental specialists.

As the process develops, the design is then often managed by the contractor's design managers who will schedule, programme and continually check and develop the design until such time as they consider it complete and ultimately approved by the client.

During this time, the design is often revised to suit budgets and new or emerging technologies that may have just become apparent or reached the market. This is sometimes driven by cost and time constraints, which often govern the feasibility and operational efficiency of the built environment See the mind map below for a summary.

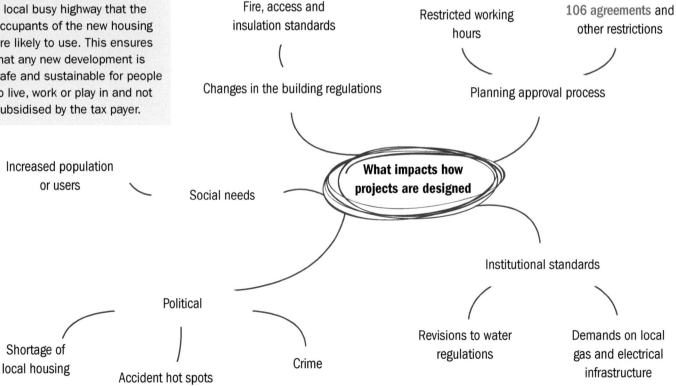

Fire, access and insulation standards

Restricted working hours

106 agreements and other restrictions

Changes in the building regulations

Planning approval process

Increased population or users

Social needs

What impacts how projects are designed

Institutional standards

Political

Revisions to water regulations

Demands on local gas and electrical infrastructure

Shortage of local housing

Accident hot spots

Crime

The design often develops due to the collaboration of specialist contractors who add value, cost-effective alternatives and practical insights to the process by using their experience of previous projects that they have been involved in. This is becoming more common as the built environment becomes more integrated, sustainable and complex. The early engagement of these specialists is now seen as vital in the design process.

The design process is often compared to the RIBA plan of works, which is a nationally recognised professional timetable of how a construction project can be managed and delivered. The plan of works breaks up the design and realisation process into stages and illustrates what and who is responsible for each stage of the design.

KNOWLEDGE // and UNDERSTANDING // Project planning

The pre-construction period, construction phase and maintenance period will be planned in detailed schedules and programmes of work at an early stage, to help inform, guide and check the procedures, activities and processes of construction and quality checking during the project.

This activity is often driven by the project manager and executed by the Programme Engineer, who uses drawings, specifications, schedules, documentation, their professional experiences and those of their team to determine a plan of work in a sequence. These programmes of work often form part of the contract and/or tender documentation, as they define strict timelines for delivery (i.e. who does what, how long for and by when).

This sequence of work is commonly produced in a variation of a long-standing planning tool called a Gantt chart. This chart presents the team with a reference document that is readily available and updated by the planner to consider changes in the procurement, design and construction process.

It is vital that all factors that influence the procurement, design and construction process are factored into this plan of work to ensure it is both current, meaningful and communicated to the entire team, to allow them adequate time to comment on the planned dates of work.

More recently, powerful software is used to produce construction programmes; however, meaningful programmes can also be produced as a result of using simple software spreadsheet-type programs such as Excel. Programmes drafted by hand are equally as useful and reliable for simpler operations or tasks. Regardless of complexity, a meaningful program will consist of the following data.

Basic information in the vertical scale (often positioned on the left side of the program):

- Type of activity (bricklaying, plastering, carpentry)
- Start date (for activity of bricklaying to start)
- End date (for activity of carpentry to finish)
- Duration of the activity (number of days or weeks for plastering).

Basic information in the horizontal scale (often positioned on the top of the programme):

- Time period in staggered lines showing year, month, week number (corresponding to month) and working days (Monday to Friday).

By regularly reviewing, developing and revising programs, while consulting stakeholders such as clients, designers, manufacturers, suppliers, tradespeople and sub-contractors, the team are better placed to perform positively and therefore the project is more likely to be a success.

KNOWLEDGE / Procurement

Procurement is another process that relates to how services provided by tradespeople, designers and sub-contractors are secured in the built environment. Procurement also relates to how goods such as bricks, steelwork, light fittings and roofing materials are ordered and secured. In the UK these materials are obtained from all over the world and delivered by land, sea and air, depending on how urgently they are required and what price the industry or a specific project is willing to pay for them.

The procurement of these construction-related services and goods is complex and not infinite, so you must take this into account when you formulate and complete your programme of work. It's fair to conclude that the procurement phase of a project will always determine the completion date of a construction project due to the availability of both services and goods.

Some organisations such as local authorities have strict rules of governance relating to procurement to ensure that suppliers have a fair and equal chance to tender for potentially lucrative and important contracts. This form of procurement is often based on both the cost and quality of services provided, so factors such as sustainability and standards of work or timely delivery are well considered prior to placing orders. For example, the photo on the left was taken at a collaborative framework meeting where the local authority, developer, legal advisor, contractor and national government met in a public forum to develop the SEWSCAP 3 framework in order to deliver best value.

KEY TERM

SEWSCAP 3: South East and Mid Wales Collaborative Construction Framework; an organisation where several local authorities and other parties form a cooperative to improve their buying power for goods and services, and to ensure best value is obtained when public money is used to construct building and infrastructure such as schools, public buildings and roads.

However, private, smaller clients may have less rigid procurement practices that are based on lowest cost alone, so this may restrict what type and size of organisation is suitable or can even compete for certain types of work, which is often based on value (£) and scope of work (sizes of contracts).

The common factors that influence procurement are:

1
Location (working in remote areas or far away from an organisation head office can be costly).

2
Value of project (a large multi-million-pound office development may be impossible for a small contractor to attempt due to restricted cash-flow).

3
Size of project (a large commercial building may be too complex for a small contractor to manage and resource without the need to employ more staff).

4
Site constraints (a multi-story office block in the heart of a city centre may have many planning conditions that require a lot of managers to ensure obligations are met during the construction phase).

5
Expectation on standards and quality (the construction of a high-quality exposed concrete may be beyond many smaller less experienced contractors' abilities).

6
Availability of resources (shortages of labour, plant or materials are common when constructing residential properties during a housing boom, so larger organisations often succeed in securing them as they are willing to pay premium prices to obtain them).

Construction (secure site, site clearance, substructure and superstructure)

KNOWLEDGE // and **UNDERSTANDING** // Secure site

The site must be secured to provide segregation of the construction activities and the public, and to prevent unauthorised entry by thieves, vandals and anyone else who is not allowed on site. A secure perimeter hoarding construction from plywood and structural timber is often used but there are many more solutions such as open-mesh fence panels (shown left), metal sheet fencing, and preparatory systems.

Many businesses take advantage of this by decorating the perimeter hoarding with their company colours or information relating to their client or the development itself. This method of securing the site is referred to as a primary security control measure and is often formed by a specialist fencing or hoarding contractor. Local police forces are also made aware of the project details and its duration, so they can help prevent crime as part of their own ant-theft and public protection campaigns.

Once completed, the hoarding or fence is then referred to as 'temporary works' by the site management, who check its condition and integrity (ensure it isn't broken) periodically during the construction phase to ensure that the wind, site activity or individuals have not damaged it.

On completion of the project, the site perimeter fence is off-hired and returned to the manufacturer or supplier. If the hoarding is constructed from timber it is often dismantled and re-used or probably sent for recycling to become another timber-based product or fuel source. This is an effective way to make best use of materials that would otherwise be thrown away or put into landfill.

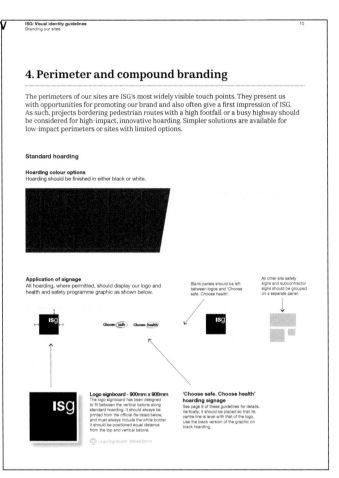

KEY TERM

Reactive guarding: The tactic of responding to the activation of a monitored automated alarm within a set timeframe. The response can be by a mobile operative attending the site but is more likely to be an audible warning demanding potential criminal and unwanted visitors to leave.

Once the site is secure from unauthorised access, additional primary measures are incorporated into the perimeter, such as controlled access gates and guardrooms and secondary measures like CCTV and reactive guarding arrangements.

Site clearance

Once the client and consultant team have permission to proceed and risk assessed the site proposed for development, the contractor is free to clear the site. The contractor will check any planning conditions that relate to pre-commencement, such as providing logistical method statements and hours of work to the local authority, and ensure this is done prior to starting. The contractor and client must also provide the Health and Safety Executive with a notification to commence work prior to starting. Existing services that may be damaged or cause accidents or disruption also must be plotted and marked out on site.

The contractor will commence site clearance only when all these pre-start checks and procedures have been made and all stakeholders are happy to commence the contract. The site clearance is a vital activity that allows elements such as vegetation, asbestos, contamination, redundant buildings and infrastructure, and waste to be removed from site in a controlled manner and taken for disposal or recycling locally.

The site clearance stage often stops when the existing ground is reduced to the underside of the ground-floor construction, so the site can be checked, tested and certified for its load-bearing potential. At this point the designer team, led by the Structural or Civil Engineer and aided by the building control officer, will determine if the site is appropriate for the next stages of the project.

They will look at factors such as further contamination, for example asbestos or oil in the subsoil, as well as the presence of tree roots, water sources and antiquities when they survey the site. Some sites that have historic significance often have an archaeologist in attendance, who inspects the subsoil at different stages of the project.

Some sites also have a procedure and 'watching brief' for unexploded ordinance, because air raids during the two World Wars may have left unexploded explosive devices just below the surface, particularly in older industrial areas that have soft or muddy ground.

Once the site is cleared, the levels of the formation will be taken by the Site Engineer and recorded, so the design and construction team have an accurate record for qualitive and quantitative purposes. When this is complete, the construction phase moves forward, often with the site being sealed with stone to act as a robust platform to carry heavy machinery, scaffolding and to act as a safe lifting area to construct the building. This also helps drain the site and prevents the transfer of muddy contamination onto the adjacent roads and carriageways.

Often the contractor will install a wheel wash device at the main gate to ensure that vehicles leaving the site are as clean as possible to mitigate pollution on the road. This ultimately helps discharge and comply with the planning conditions that the local authority may have imposed on the development.

Handover to client (commissioning and handover)

The building can only be handed over to the client and deemed as complete when it is fully completed and all its electrical, mechanical and data installations are tested, working properly and have been certified by a commissioning engineer.

This is because the systems within a building, such as the lighting, wiring, plumbing and communications, are often complex and rely on utilities such as electricity, water, gas and fibre-optic networks connected to a wider national grid to perform properly. This means the systems within the building must perform to rigorous standards and approved codes of practice.

The building structure and fabric must also be tested and inspected by the Designers and building control officer to ensure that elements such as fire integrity, means of escape, and access for wheelchair users and vulnerable users are accommodated for within the design, constructed as part of the works on site and working as they were designed to.

All planning conditions imposed by the planning officer must be discharged and agreed as completed before any development can be considered complete.

These requirements are only proved to be completed when those who are responsible issue vital certification and written confirmation that these elements are indeed complete. Once these are received, they are condensed into the building's documentation in the form of a building log book, health and safety file and operating and maintenance manuals (just like you would get with a new car). The production of these documents and certification are the product of collaboration between the client, designers, contractor and specialist sub-contractors that has taken place for many weeks, sometimes many months beforehand.

LINK

For more on Soft Landings see page 168.

More recently, this process has become known as the Soft Landings, which is an initiative driven by the government to achieve ultimate performance of the building and its systems, going well beyond the handover stage and many years into its designed life.

This ensures that the building performs as it should and optimises its energy consumption with features such as regular reviews of energy consumption, sub-metering and regular maintenance reviews. It also ensures that the building users and owners, who may change many times over the years, know and understand the features of the building, its systems and important facts such as where to obtain spares and replacement components like valves, fuses, switches and appliances, or what components are structural if they want to form a new opening within the structure or they just want to simply clean the roof or gutters safely.

Maintenance

Maintenance is the actual process of planning and doing vital work on a completed building or component or both. As buildings become more complex in their form, function and operating systems, the need for more reactive, planned and preventative maintenance becomes greater with it.

Even the most basic activities of cleaning and lubricating mechanical components (locks, valves, hinges and mechanisms) and building fabric (walls, floors, windows, glass, gutters and roofs) can have huge positive impacts on performance. The primary objective of maintenance is to ensure the optimum performance is achieved and that decay and derogation is minimised.

This activity is done by directly employed members of staff and/or professional maintenance organisations, who are made up of multi-disciplined teams of specialist construction, mechanical, electrical and data specialists. These teams of construction professionals regularly survey the building and its systems to an agreed programme of maintenance agreed with the landlord, building owner or building user.

These organisations have full access to the building, its systems and handover documentation, and operate to a set of pre-agreed schedule of rates to ensure all the systems are maintained. These teams also replace defective elements and components as well as maintain and repair them.

As maintaining the built environment becomes more vital to keep it operating to its best capability, the need for meaningful and relevant handover information has led to the Soft Landings process, which starts just as the construction phase commences, ensuring that maintenance of the building is also high on the agenda of designers and builders alike. This makes the maintenance data and processes more attainable and practical for those that both build and maintain the built environment.

AC2.2 Calculate resources to meet requirements for built environment development projects

The **calculation of resources** that are required to deliver certainty relating to quantities, cost, quality and standards within the industry is a vital set of skills that all stakeholders need in order to function on a daily, hourly and sometimes minute-by-minute basis. These outputs help the industry to communicate quantities of materials, expectations on quality and standards of work, and allow you to have a numerical language that everyone can understand and relate to.

This allows a diverse industry of developers, planners, clients, tradespeople, sub-contractors and construction professionals to agree on the acceptable limits of what they expect in return for payment or must provide because of entering into a form of contract or part of an agreement.

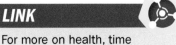

LINK

For more on health, time and resources see page 31.

UNDERSTANDING Recalling, once again, your three precious things of time, health and resources, you can immediately observe that **resources** remains one of only three things common to everyone. Therefore, you must preserve the World's resources and ensure they are utilised in a sustainable manner, particularly within the built environment where resources remain in high demand and appear to be diminishing rapidly on a global scale.

SKILLS Calculation of the following factors helps to identify positive and negative aspects of many things that relate to your understanding of the figures, sums or values of the resources you are attempting to order, purchase, distribute or utilise.

These figures, sums and values help you to make decisions based on numerical facts, and often promote more questions relating to the resources (labour, plant and materials). There is an infinite number of questions and answers relating to calculations but here are some common ones followed by explanations and reasoning relating to them.

APPLICATION Area

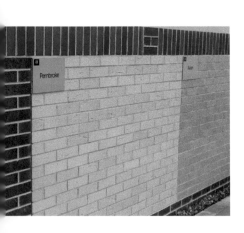

How many bricks do I need to order to ensure I have enough materials on site to complete the brickwork in five days?

In the construction industry, 'area' generally relates to the actual surface area of materials such as brickwork, plaster, decoration, glass and even the surface of the tarmac or the ground itself. This measure is a common way of rapidly quantifying and agreeing how much work has been completed by the tradesperson and how much it is worth in terms of value to the Quantity Surveyor. Many trades are often paid on a m^2 rate. For example, a bricklayer may be paid £10.00 per m^2 for their work. If that tradesperson completes $50m^2$ of work per day and works for five days, then they will apply for a payment from the Quantity Surveyor of £2,500.00.

APPLICATION / Volume

How much water will I need to fill the newly constructed swimming pool?

Volume is commonly used in the construction industry to quantify the amount of a substance or material contained within itself or a container. There are many instances where volume calculations are used such as the measurement of concrete, water and loose stone. The great thing about volume is that it can be readily converted from m^3 into kilogrammes or litres, which helps you to translate and understand the perceived size or weight of the volume of the material or substance being considered. For example, a swimming pool that is 20m long × 10m wide × 1.400m deep will need $280m^3$ of water to fill it. By converting the $280m^3$ to litres and kilogrammes you build a greater comprehension of the significance the water has on the design, installation and maintenance:

Formula	m^3	Litres	Kilogrammes
Length × width × depth	$280m^3$		
Multiply the volume × 1,000		280,000 litres	

Note: The calculations in this table are for water only.

APPLICATION / Percentages

What percentage of the newly plastered wall needs repairing so I only pay for compliant and acceptable standards of work?

Percentages are a common means of understanding the specific part or composite part of a material that has been formed into products or requires some form of division. This is an effective way to ensure a fair share of the reward or responsibilities in question. For example, a newly formed plastered wall of $30m^2$ has been recently damaged by a decorator who has scratched it with their mobile platform. A meeting has been held on site and the plasterer, site manager and painter agreed how many m^2 of plasterwork is damaged. They mark up the damaged areas and the Quantity Surveyor measures the area and concludes that it is $15m^2$:

$15m^2$ divided by $30m^2$ multiplied by 100 = 50%

APPLICATION / Scaling

What scale shall I use to represent this large site as I want to make sure the plans are legible?

Scaling is an important tool that is used in the construction industry to help you comprehend the ratios of measurements to a consistent scale. The scale is often agreed at a scale that is comfortable to the viewer's (person's) eye. Some common scales for site plans are set at:

- **1:50**: for every 50mm of actual length, this is equal to 1mm on the scale rule.
- **1:100**: for every 100mm of actual length, this is equal to 1mm on the scale rule.
- **1:200**: for every 200mm of actual length, this is equal to 1mm on the scale rule.

The scale can be found on both hard copies and digitally issued drawings, usually on the lower right-hand side of the drawing. Always check what scale you are reading every time you pick up your ruler, as you can easily turn it over to a different scale and therefore may give you a value different from that on the drawing.

Larger scaled drawings greater than 1:500 mean that the site is really large so needs to be condensed down in scale to fit it onto the paper or screen. Scales of 1:5, 1:10 and 1:20 indicate that detail represents larger aspects such as a section through the building showing foundation, substructure and roof details.

 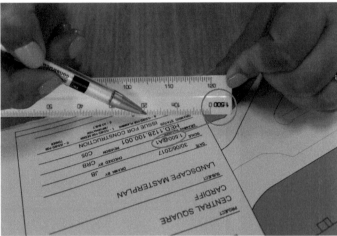

APPLICATION / Best value

How can I explain what best value is to our client as they might think that our rates are excessive for the refurbishment of the ground floor of their house? After all, they will be living in the upstairs part of the house while we work so we will need to be considerate contractors and the work will take longer.

Best value is a term that the UK's government introduced in the mid-1990s to raise awareness of the poor relationship that both local and national governments and the public sector had with industry when it came to purchasing goods or services. Traditionally, industry provided goods such as light fittings and taps to the schools and hospitals that were low cost and had short life spans, which were sourced to suit their restricted budgets.

This method of procurement gave rise to cheaper and cheaper products of poor quality that had a very short life span. This meant schools, hospitals and even the military were spending huge sums of money replacing building components that were not operating efficiently.

Best value is the procurement process by which the buyer investigates the value of these goods and services not just by its commercial cost/value (£) alone.

The following factors are also considered by the buyer to help them conclude the value of products and services:

- Was it made locally or in a country where child labour is used?
- Is it made from any recycled materials?
- Do the contractors who tender for the work have experience of constructing similar projects?
- Does the manufacturer provide a guarantee with the product?

- What happens if the boiler breaks down after one year?
- Is the specialist sub-contractor local or will I have to wait long periods of time if there is an issue after completion of the project?

As the saying goes, 'buy cheap, buy twice'. Best value can only be decided by those who are purchasing the goods and services, as they know what's important to them. Best value procurement has become so popular that both public and private sector clients follow this route of procurement to optimise their purchasing power to get best value for their money. When appointing a contractor for best value in the above scenario, the home owner could focus on questions such as:

- How many times a day will the contractor clean the works?
- Will the contractor work within restricted times of day and weekends?
- Has the contractor made allowance for dust extraction to ensure a healthy shared environment?
- Has the contractor made allowance for temporary windows and doors to maintain security?
- Will the contractor be paying for any electricity or water they consume?
- Does the contractor have any experience of refurbishments?
- Will the contractor provide their own WC and canteen or do they expect to use my home?

APPLICATION / Tolerances

We need to agree the tolerances for setting out the new extension, particularly as the existing house is 65mm out of square to the boundary wall. There are also two different floors levels on the existing house, so which one do we use, or shall we calculate an average of the two?

The construction industry relies upon lots of different people with different skill sets and abilities to construct buildings and everything that they contain. This means often using products that are assembled off and on site in lots of different climates and under varying conditions. Different parts of the building will often contain different types of materials, each with their own properties and characteristics.

For example, steel, timber, glass and concrete are commonly used together and form junctions with one another, yet they all perform in different ways. If they are exposed to the elements during the construction process, they may therefore react differently when they dry out. This could lead to movement and cracking, which is not desirable. When these products are installed to current standards and **abut** an existing house that has inconsistent shape and floor levels, then there needs to be adequate and reasonable tolerances.

The design for the extension above needs to accommodate the new materials so that the structure is sound, the elements are kept out of the building and the floors are flat where new and existing meet, so the occupants don't trip over. Tolerances are the primary quality control measure used to set the standards for the design and construction process, and give a definitive value that the tradesperson must work to when constructing the extension.

If the correct materials are selected to cope with the irregular shape of the existing building and the details formed correctly and within tolerance, then the project is likely to be a success.

LINK

For more on tolerances see page 82.

KEY TERM

Abut: For example, where a building, part of a building or boundary of land meet with one another. 'The external north elevation of the farmhouse abutted the public footpath, so the owner decided that they wanted to use mirrored glass within the window frames to prevent pedestrians seeing into the kitchen.'

APPLICATION / Value added tax (VAT)

How can I work out how much VAT to pay on the work relating to the carpentry work on the five new houses that we have just completed?

Most businesses in the construction industry must pay VAT (value added tax) if they earn over a certain amount. This tax is a percentage (at time of writing 20%) added to a product or service at every point where value is added. This means that if you are a carpenter and you provide and fix materials such as doors and ironmongery on a housing project, then you add value to the completed work. This also means that you will likely charge VAT to the company or person you fitted them for. Meaning that you will add 20% to your total fee because you have added value by incorporating them into the project. This tax is then collected by the government at the end of the tax year.

APPLICATION / Tender price

How can I explain to Mr Jones that the tender price is different from the proposed final account figure that we need to rely on? I hope there's adequate reserves to pay for all these additional works he instructed!

A tender price is a response from a contractor to a potential client to an invitation to tender for work. This tender price is a defined budget that lists out (often in detail) what the contractor has interpreted as being required to complete the works. The tender price is often summarised into the main elements of the project such as substructure, superstructure, envelope, finishes, etc.

These elements will have the anticipated costs and quantities that the contractor has obtained by studying the contract documents, schedules and drawings. Larger organisations obtain quotations from sub-contractors and other specialists to ensure that the tender price is compliant to the client's needs and documentation.

For example, Mr Jones received a tender price for the construction of a new bungalow that had a total cost of £250,000.00. During the construction phase Mr Jones changed the specification of the bricks and layout to the interior by issuing instructions to the contractor. The contractor must now provide Mr Jones with a summary of the actual costs for the work which has increased, as the materials have become more expensive and the extent of work has changed from that of the tender cost. Therefore, as some costs may rise and some may fall, an accurate cost of the work must be submitted to Mr Jones for him to consider.

APPLICATION / Labour, plant and materials

The association of labour, plant and materials to the calculation of all these factors is clearly related to the 'resources' (one of our three precious things). Consider how the comprehension of these elements interacts with one another as the proportions of labour, plant and materials all vary in relation to one another. Research online how much a fair-faced brickwork wall of 20m long and 1m high currently costs to construct at current labour, plant and material rates.

Conclude what the costliest element per $1m^2$ of brick wall is valued at, is it labour or plant or materials?

AC2.3 Access potential effect of factors on project success

In the UK and elsewhere in the developed world, the construction industry and the industries that help it function such as haulage suppliers and financiers form a substantial percentage of everything produced. In the UK this percentage varies depending on what source you obtain your information from, but the general consensus is that the industry generates about 8%–11% of everything that is produced. This is known as gross domestic product (GDP). To have such a large part of everything produced related to construction is both a risk and an opportunity for the government and those that work within the construction industry alike. This is because when the economy is strong, investors such as homeowners, developers and those with capital (£) may want to use their money to build and refurbish building stock, as this adds value and creates wealth very quickly. The result of this creates a positive effect on the economy and helps businesses to grow, creating greater confidence in the industry. In turn, people feel happier, more financially secure and the positive effects help the economy to grow further.

UNDERSTANDING / In broad terms there are three reasons why there are ups and downs in the construction industry that we need to understand for this unit.

Reason 1: Cause

When the economy is not so prosperous, the confidence of investors is not as strong because they feel less financially secure, so they keep their money relatively safe in other types of savings and investments such as high-interest bank/building society accounts or foreign investments such as oil, gas and manufacturing. Or they spend it on essentials such as food, fuel bills and clothing, or luxuries like holidays or a new car.

Reason 2: Effect

The low demand creates pressure on the construction industry as it does not have adequate reserves of labour, plant and materials to cope with rapid changes or sudden demands. This is not only due to the factors in 1 above but also because there are insufficient managers, bricklayers, carpenters, etc. in the trade. Factors such as changes to construction-related laws and regulations, and advancements in technology like 3D printing and BIM, also pressurise the industry to adapt and upskill.

Planning governance, shortage of development land and the impact of new and emerging industries associated with the digital age also add to the shortage of tradespeople by the workforce being attracted away from the industry.

155

Reason 3: The economy

As the economy rises and falls, the confidence of those who live and work within it have their spending decisions influenced by whatever the good, bad or indifferent feelings are at that point in time. These rises and falls in confidence can sometimes last weeks, months or years, so the construction industry always must grow and shrink with the economic environment at that time.

This is one of the main reasons why the construction industry can be so lucrative and successful but can rapidly become unprofitable and unsustainable.

Regardless of the positive, negative or even indifferent condition of the economy, there are both internal and external issues that influence the factors of success on how projects are realised that remain consistently challenging to the industry.

KNOWLEDGE / and UNDERSTANDING / Internal factors

Finance

Funding for many construction projects rarely comes from an individual's or developer's savings or reserves. Most construction projects realised in the UK are financed by a multitude of savings, loans and financial products. Securing finance for developing any size of project will depend on many factors, including the individual's or organisation's credit rating, experience, knowledge and the viability of the building project itself.

Qualified workforce

The UK construction industry has an enviable worldwide reputation for its standards and expertise, which demands the very best managers, designers, tradespeople and support staff. Having a workforce that is qualified to these exacting and high standards is both expensive and takes time to achieve. Specialist occupations, processes and rapidly changing methods of manufacturing, designing and constructing the built environment mean that the qualifications need to keep pace with the industry.

Certified workforce

Similarly, the UK has an equally enviable global reputation for producing tradespeople, sub-contractors, specialist contractors and supply chain members. This essential workforce population must keep fully certified and accredited to utilise traditional and modern methods of construction. Health-, safety- and environmental-related certification is vital to maintain a strong workforce and safer construction environments.

Operative and supervisory certification for using specialised plant, driving vehicles, working in confined spaces and working at height all demand high standards of training. The positive effects of this help keep the industry sustainable and drive the safety agenda to ensure workplace accidents are minimised and occupational health is improved.

Security

The security of our built environment is vital to keep the viability and success of construction projects. Organised crime groups, vandalism and theft cost the industry millions of pounds in lost revenue and production every year. Unfortunately, smaller organisations often fail to invest in adequate security measures and fall foul of these criminal groups.

This site security is not confined to crime alone, as there are many accidents and incidents indirectly impacting the public which are caused by poorly secured sites. This is not only illegal but totally avoidable given the multitude of cost-effective security solutions available on the market. Many such incidents are often not covered by insurance policies, which highlights the need for employers and employees to be responsible for security, both morally and legally.

KNOWLEDGE / and UNDERSTANDING / External factors

Penalty clauses (now referred to as damages)

Penalty clauses (now referred to as damages) relate to mechanisms within building contracts that entitle the client to what are commonly known as liquidated and ascertained damages (LADs).

This type of penalty clause is often used by the client to recover costs for lost revenue and profits they would have otherwise made from selling or leasing the building, or if a domestic client intended to use it for a home, for example, then the cost for living in alternative accommodation or a hotel would be recovered from the contractor, through the contract.

The level of the damages will be agreed as part of the contract and defined at a daily rate, for example:

> *The liquidated and ascertained damages for the project will be set at £500.00 per/day for every calendar day that the works are deemed not practically complete.*

These associated costs may also relate to covering the client's rent, storage, loss of income, fees and interest on loans.

This gives the contractor and client surety that there are reasonable and pre-determined 'costs' for failing to deliver the project, but only if the client has given instructions for additional works or failed to provide instruction or access to areas, at which point the contractor is also entitled to recover costs.

Weather

The climate and weather patterns in the UK are difficult to predict and are changeable over a very short period of time. This is because of the UK's position in relation to the jet stream, North Atlantic Ocean and North Sea, and the diverse topography (features) of the UK's mainland. This means that wind, driving rain, frost, snow and sometimes intense heat can disrupt construction activity on site and delay construction programmes and methods of working.

For example, the construction of brickwork in winter in a damp or cold environment can lead to efflorescence and cracking, similarly concrete work spoilt by frost can

KEY TERM

Efflorescence: When fine salt-like chemicals migrate (move) to the surface of materials such as brickwork and render, often using rainwater as a vehicle. The result of this migration is white powder residue that is left behind as the drying-out process occurs. In severe cases, solid white salt-like deposits become crusty and solidify, which are very unsightly and difficult to remove.

Efflorescence on concrete parking tiles.

be financially harm a company if large sections of concrete must be re-formed if it is not protected with frost blankets. In warmer periods the sun can prevent industrial roofing work, as the glare and radiation can damage workers' eyesight and burn their skin; rapid drying can also irreversibly damage concrete, floor screeding and external wall rendering.

SKILLS / Copy the following table into your notebook and then complete it by identifying the risks (how could the weather damage unprotected work) to activities in both summer and winter.

Activity	Risks in summer working	Risks in winter working
Forming external concrete floor slabs		
Constructing south-facing brickwork walls		
Placing industrial roof sheets		
Forming landscaped gardens		
Drystone walling in a rural and exposed area in Northern Scotland		
Forming a stone slate roof on a cottage in Cornwall		

Research online how these activities can be protected from the elements when working.

AC2.4 Interpret sources of information

The interpretation of sources of information allows you to empower yourself with information and knowledge. Academics, politicians and business leaders have a common saying and some even base their whole careers on it:

INFORMATION IS POWER.

KNOWLEDGE This concept is of course fatally flawed, as history shows that an individual can have lots of information, but for it to be of value it helps if it is:

1 True, factual and accurate

The information is based on accepted principles that can be proven by research and development programmes, testing regimes, certification and evidence that it has been successfully applied before. In the construction industry many sources of information (data) are used to define if methods such as an external render system, products and processes such as a waterproof paint and dry-rot products and design solutions such as the layout of a new kitchen are logical and safe to use.

2 Current and relevant

The source of information must be current and relevant. For example, a design proposal from an architect for a new stair must comply with the latest revision of building regulation requirements for stairways or a proposed type of paint system must have a BBA (type of testing certificate) that is recent to show that the contents currently used within it have low volatile (changes rapidly) chemicals content.

Furthermore, the information should be relevant as it must be capable of becoming a compliant solution in the first instance because there is little point in considering a non-compliant product, design solution or process if you later find it to be detrimental to a task, such as the use of untreated softwood in an external environment or a type of paint that can't be washed down has been applied to a wall adjacent to a kitchen worktop or domestic food preparation area.

3 Applied in a hierarchy and approved by the client (or their representative)

Information that is used to underpin and conclude if a design solution or product that you propose to be used is compliant to the **specification** (often aligned to accepted British Standards, approved codes of practice and/or the experiences and knowledge of the client's design team).

All proposed design solutions and products must then be approved by the client and the design team to check that it is fit for purpose.

By applying these skills of investigation and checking compliance criteria and feasibility you can use the following sources of information in the correct order to help process the information in the most efficient way:

1 **Specifications**

2 **Drawings**

3 **Spreadsheets**

4 **Catalogues**

5 **Suppliers' materials lists**

UNDERSTANDING // Sources of information can be referred to as primary and secondary sources of information.

- **Primary sources** of information are sources that are known to be objective and to be used as the true basis of comparison. It is your objective to satisfy the criteria within them and provide compliant design solutions and therefore satisfy the wants and needs of the client and their design team.

- **Secondary sources** of information are sources that have been supplied sometimes as part of the contract information (subjective) but generally from suppliers of aggregates, cement, paint and so on; contractors such as bricklayers, plasters and carpenters; specialist contractors such as window and glazing technicians; or the wider supply chain that produces products including furniture, lighting and flooring, or sells services such as interior designers, upholsterers, acoustics specialists and audio/visual equipment.

SKILLS // **Primary sources of information**

Specifications

These will vary from project to project and are intended to be descriptive documents that define workmanship and materials. There are two types of construction-related specifications and these are:

1 Performance specifications which are used when a building or building component requires designing such as a new home of multiple occupancy for the elderly that demands additional access or security measures, or a component like a specialist low-pitched membrane roof that has to be a special colour for planning considerations.

220 Site visit

- **Nature of the site**: Ascertain before Tendering, including access thereto and local conditions and restrictions likely to affect the execution of the Work.
- **Arrangements for visit**: Telephone to arrange appointment

230 Return of Tender

- **Return of Tender**:
  - **Destination**: Architects Office
  - **Time and date**: 12.00 noon, Friday 16 June 2017
  - **Format**: As described in the invitation to tender
  - **Special procedures**: None
- **Documents to be returned with the Tender**: As described in the invitation to tender and the Specification.
- **Inability to tender**: Advise immediately if the work as defined in the Tender documents cannot be tendered.
  Define those parts, stating reasons for the inability to tender.

310 Assessment

- **Assessment of Tenders**:
  - **Number to be assessed in detail**: All.
  - **Assessment criteria**: Most economically advantageous.
  - **Assessment model details**: As described in the invitation to tender
- **Alternative Tenders**:
  - **Submission**: Not permitted.
  - **Basis**: N/A

320 Error resolution

- **Arithmetical errors**: Tender price will prevail. An opportunity will be given to confirm the Tender or withdraw.
- **Technical errors**: The Tender is deemed to meet or exceed the requirements of the Tender documents. Amendment of the

Tender to reflect this will not constitute a variation and no claim for additional costs will be accepted.
- **Corrections**: An endorsement will be added to the priced documents indicating that rates or prices (excluding preliminaries, contingencies, Prime cost and Provisional sums) inserted therein will be adjusted in the same proportion as the corrected total differs from that stated incorrectly.

340 Post-Tender negotiations

- **Details**:

410 Notification to Tenderers

- **Notification method**: As described in the invitation to tender

NBS Create Sample Specification – Project Management

14

An extract from 'NBS Create Sample Specification – Project Management' (2018).

2 Prescriptive specifications breakdown a very detailed description of how something like a specific product, such as a handmade clay brick, should be used on an elevation of a building in a conservation area, or a specially selected tree species to help regenerate native woodland. The NBS is an organisation that unifies the production of specifications to have a common way in specifying most things for many of the industries. It considers a specification is best written and understood if the **7Cs** method is applied, as shown on the right:

1 CLEAR

2 CORRECT

3 CONCISE

4 CONSISTENT

5 COMPLETE

6 COMPREHENSIVE

7 COORDINATED

Drawings

The majority of construction drawings issued and read today are generated by powerful digital computer programs that make accuracy, scale, content and sometimes conflicts in the data (data that doesn't add up or agree) less likely to occur. This means the drawings are more likely to be of greater accuracy than ever before. When a drawing is received the reader should check the drawing title table that is generally located in the lower right-hand side of the drawing to check the validity of the drawing (that it is the correct one).

This table will show important data that is revised or updated during the design development process until such time as it is ready to be used for the construction phase of a project. This may be revised during the construction process and then it will be revised for the final time just before handover to the client to show how the building was constructed. The client may then revise it further as the use, age and condition of the building changes through its lifecycle.

Every drawing must come with a legend or key to allow the reader to understand the meaning of the symbols within. This is because different designers often use different symbols to reflect the information such as brickwork, blockwork, electrical sockets and locations of doors.

When interpreting a drawing on site it is often helpful to orientate it to the same direction in which you are facing. Similarly, if you are comparing two drawings then they too should be orientated the same way to each other to avoid confusion.

APPLICATION / Research the symbols online so that you can start to understand how pictograms vary but generally have the same meaning. Draw a simple plan of a room familiar to you then face in different directions, orientating the drawing to suit the way you are facing.

SKILLS / Drawings have three primary stages of development that take them from a concept in someone's mind to a scaled pictorial representation or digital model. These are:

1 Preliminary

These are conceptual drawings often drawn by hand then reworked digitally into plans, elevations and sections that show the basic shape of the site, its topography and any constraints such as boundaries or roads. They are created by designers for the client to help them develop ideas and collaboratively move the process onto the next stage. This helps both the client and designer get the best ideas as early in the process as possible and saves valuable time. When revised they are often followed by the suffix P1, P2, P3, etc. until such time as they are ready for the tender process.

2 Tender drawings

These drawings now show far more detail and the client has approved them for the tender process. They have enough detail on them to allow the client to obtain several quotations from building contractors.

During this time the designers have a dialogue with the client and builders, ensuring they keep a schedule of any queries asked by the builders in case they have missed any features or even to capture positive suggestions on how to improve the design. Any revisions made by the designers are prefixed with 'T' so the team can reference them when costs change or features are revised, such as changing the shape of a floor plan to accommodate an additional door or window, as this will have an additional cost. Safety considerations are also developed to ensure that CDM regulations are adhered to.

3 Construction drawings

The successful contractor has been appointed and the designers now issue construction drawings that accurately reflect what the client wants and their needs. These drawings are ready to be used by the building contractors to construct the building. Any changes during this time will result in the drawings being revised, with the suffix 'C' used to help keep track of the design.

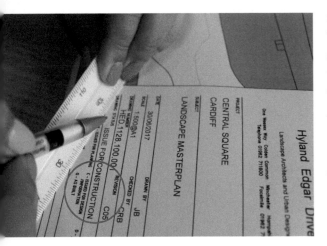

Status of drawings

The Architect, who is authorised to give final approval of the drawings after checking that the client or user of the building is satisfied with the design, will relay their conclusion using a suffix in the status box within the table at the lower right-hand side of the drawing.

This allows the team to understand that the drawing has been developed, checked and double-checked by everyone who is responsible for designing and coordinating the drawings. This means that often only the **lead designer** can have the authority to take this action after checking that the client and other design team members, such as the Structural, Civil, Mechanical and Electrical Engineers, are content that the design is compliant.

- Status A is the highest level of drawing approval and means that the reviewer can find no issues with the design.

- Status B is the moderate level of approval and indicates that there are minor revisions such as spelling or wording still to be made, but the drawing can be used for manufacturing or construction purposes, providing any recommendations are followed.

- Status C means that the drawing is not in a condition to be used for construction or manufacturing purposes.

Why are there so many levels of approval?

Several levels of approval are needed to make sure that the complex design process is compliant. This is because any mistakes at this stage will be very expensive in terms of money and/or time. As drawings can take several days or weeks to revise and update, the status helps the builders know if they can order materials and therefore save time on the programme (as long as they consider any comments made by the designer on the drawing).

Secondary sources of information

Spreadsheets

Spreadsheets are an effective way to format and process large amounts of data such as dimensions, areas, time and cost. Spreadsheets can be shared to save time, calculate values automatically, improve the accuracy of cost and quality-related data, and can be easily and readily revised, updated and shared again.

Spreadsheets are used by the entire industry to record most numerical data such as costs, lengths, quantities and common features of products or objective-based data such as accidents, incidents, dates, times and colours. They can express data in charts and diagrams automatically if used correctly and have almost infinite potential to help the industry quickly compare, contrast, filter and collaborate.

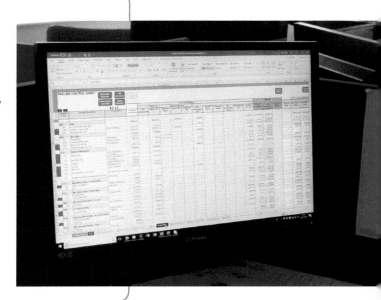

Catalogues

Catalogues are effective ways for manufacturers and suppliers of design solutions, building systems and construction-related products and materials to reach the marketplace. Catalogues can be printed or digital documents that are linked to data-rich environments where they can be further promoted on channels such as video-sharing platforms and also on the website of those who have used them to great effect such as specialist timber-frame companies that can erect homes in only a few days, all the way through to paint products and furniture. These documents and data-rich promotional materials are also used by the industry to compare, contrast and specify proposed design solutions.

Suppliers materials lists

As technologies advance rapidly, the change within the construction industry on the range of materials and processes has affected how we design and construct the built environment. The choice and pedigree of products is vast.

Best value is often at the forefront of product and process selection to ensure high quality, fairest cost, and sustainable and ethical sourcing of materials. Suppliers identify this early in the procurement process and so regularly review their manufacturing and distribution systems to remain competitive. Suppliers' materials lists are an effective and efficient way to compare current and compliant available products.

The Forest Stewardship Council (FSC) is an international organisation that operates a forest product certification scheme to ensure that forests are responsibly managed and the products from those forests can be tracked through the supply chain.

These lists can be generated by an organisation internally and used as a restricted list of suppliers. There are also several web-based lists available in the form of databases that the industry share and use.

Good examples of these lists are the RIBA product selector (https://www.ribaproductselector.com/) and the Barbour product selector (https://www.barbourproductsearch.info/#).

The RIBA product selector.

The Barbour product selector.

APPLICATION / Use these digital tools to research and specify building components that specifically may be of interest to you.

APPLICATION / During your course keep a digital record of products relating to your coursework and homework to act as a quick reference. Keep this digital record and use it to construct your own spreadsheet of associated costs for the products that you have selected.

LO3 Be able to plan built environment development projects

AC3.1 Sequence processes to be followed

AC3.2 Apportion time to processes

AC3.3 Set project tolerances

Planning forms a critical, detailed and lengthy process of **organising the construction process into a logical and realistic period**. This period of time is known as the construction project lifecycle and is not to be confused with the lifecycle of the building.

KNOWLEDGE / ## The process

The planning process is completed by specialist construction professionals called Planning Engineers or senior members of the operations team who have experience of many kinds of different projects. These construction professionals will coordinate and collaborate with the whole team, using all the documents, specifications, schedules, drawings and their past experiences.

The conclusions from this process will be captured in a **schedule of work** and reflected in a **programme of work**. This programme will capture and present **the entire process and define an overall time period for it**. Finally, this period will be presented to the directors and estimation team who will place a value on it in readiness for a total cost of the works to give to the developer. **This is an important part of the construction process and is often the decisive factor in a contractor winning the project or not.**

KEY TERM

Construction project lifecycle: The length of time it will take to plan, construct and handover the building project.

Lifecycle of the building: The period of time from when the building is completed to the end of its useful life, which usually is followed with its demolition and recycling.

UNDERSTANDING / ## Planning phase

This is the time allowance for the pre-construction team of designers, estimators, planners and builders to receive the tender information of schedules, specifications, drawings and contract data, to then translate, read and understand it. The team will meet regularly and decide the safest and most efficient way to construct the project, and will produce detailed plans, method statements and coordinated safe systems of work. During this time, they will liaise will stakeholders such as the local authority and discuss any risks like the impact of the work on public footpaths and highways, and placing plant such as skips on the highway.

The team will apply for any licences such as footpath closures, scaffold erection or site hoarding that may need specific lighting or signs to help protect the public. They will also visit the site and see how they will access and egress the site and how any planning conditions might affect the project while it is under construction, such as restricted working hours, temporary traffic lights provision or impact on local shops' or premises' deliveries.

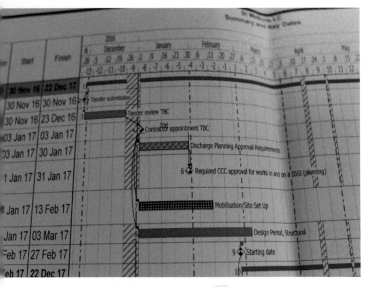

The estimators will be in constant dialogue with the potential client's team, tradespeople, designers, specialist sub-contractors and the wider supply chain to clarify any queries in relation to the tender and obtaining quotations for material and work packages. The Planning Engineer will focus on the anticipated resources and proposed construction time periods for these packages, looking in detail at the time needed to manufacture elements such as steel frames, obtaining large quantities of bricks and anything else manufactured off site like timber frames/roof trusses and specialist joinery.

Once the team understand all these requirements, a time period of attending to them, together with a cost for providing them, will be agreed and factored into the overall cost (preliminaries, labour, plant, materials and overheads, and profit).

KEY TERM

Concurrent: Activities that can happen at the same time, for example bricklaying and roofing.

Construction phase

The time allowance for this period is likely to be apportioned with the most complex and concurrent series of activities and work. These are often broken down into sub-sections of work, for example a traditionally slated roof element will be broken down into:

1 Fix roof membrane/felt

2 Setting out batten centres

3 Fix battens

4 Check fixings

5 Load out slates

6 Fix slates, etc.

The construction phase often allows for any mobilisation period that is required for the contractor to prepare to take over the site or any other site investigation works that may need to happen, such as additional structural surveys or asbestos investigations that may have arisen as a result of queries raised in the tender process.

Once again, the Planning Engineer and operations and commercial specialists will work in collaboration to ensure the best possible start by placing orders with key sub-contractors that will be working early on, such as the groundworker who will form the foundations and the carpentry contractors who can commence making the roof trusses off site to make the installation as safe and quick as possible. All these marginal gains of saved time can add up to several weeks' reduction in the overall time of the construction phase.

This is where having a professional and experienced delivery team and committed supply chain makes all the difference to a successful building contractor. More recently, modern methods of construction such as the use of off-site fabricated wall panels have re-emerged to reduce the overall construction time and help mitigate shortages of skilled trades like bricklayers and plasterers.

Handover phase

The handover phase of the project is vital to ensure that the smoothest possible transition from the construction phase to the in-use phase of the building project is realised. Many contractors are judged by the client on this phase alone, so, even if the tender and early part of the construction has gone very positively, if the handover phase isn't well planned and executed and the client is uncomfortable or has a bad experience, then this often is the difference between **repeat business or not**.

Here the builder must see the project from the point of view of the client, so certainty of the completion date, agreed final accounts and the needs of any of the client's tenants or users are crucial to handover, including things such as further fit-out projects for commercial premises, storage and removal arrangements of new home buyers, and logistics of relocating business and family possessions into the newly completed buildings. Insurance arrangements and meter readings must also be considered and managed in readiness for handover.

The building contractor must also ensure they produce the correct test and commissioning certification, that any building control and planning conditions are discharged, together with any handover documentation such as operating and maintenance manuals and a completed health and safety file. This ensures the client and the user understand how the building operates. An important part of this procedure is providing training and demonstrations to the client team on all the systems of the new building, so this too needs managing. Most larger construction projects in the UK now commence planning for handover in week 1 of the construction phase and follow the Government Soft Landings (GSL) process.

SOFT LANDINGS™

SKILLS The vital skill of recoding the sequences, processes and time allowances is a fundamental part of the Planning Engineer's function. The technique of work breakdown is a skill that is transferrable to any other occupation where detailed, intelligent and demonstrable planning is required.

APPLICATION The example of a project work breakdown technique is shown on the next page. Draft your own spreadsheet and start working from left to right to capture the work breakdown.

Copy the format of the check list on the next page and develop it for yourself.

KEY TERM

Government Soft Landings (GSL): Soft Landings is a building delivery process which runs through the project, from inception to completion and beyond, to ensure all decisions made during the project are based on improving operational performance of the building and meeting the client's expectations. (Source: BSRIA, 2018, https://www.bsria.co.uk/services/design/soft-landings/)

Work breakdown: Construction of 24 Palace Avenue			
Project name	Element	Breakdown of element	No. of days
New house construction at 24 Palace Avenue	1 Drainage	1.1 Internal drainage	4
		1.2 External drainage	3
	2 Foundations	2.1 Internal foundations	5
		2.1 External foundations	5
	3 External walls	3.1 Blockwork	12.5
		3.2 Brickwork	1.5
	4 Floor slab	4.1 Prepare sub-base	1
		4.2 Lay membrane	1
		4.3 Pour concrete	1
	5 Internal walls	5.1 Set out studs	4
		5.2 Fix sole plates	5
		5.3 Fix head restraints	5
		5.4 Fix vertical studs	3
		5.5 Fix noggin and pattress	3
	6 Roof structure		
	7 Roof structure		
	8 Roof coverings		
	9 External doors and windows		
	10 Gutters and downpipes		
	11 Watertight milestone		
	12 M&E first fix		
	13 Carpentry first fix		
	14 Plasterboard walls		
	15 Plastering		
	16 Carpentry second fix		
	17 M&E test and commission		
	18 Decoration		
	19 Landscaping		
	20 Final inspections and snagging		
	21 Clean house		
	22 Handover files ready		
	23 Demonstrations to client		
	24 Handover		
	25 Depart from site		
	Running total of days required to build new house		54

Work progressively from left to right starting with the project, and adding columns as you proceed; the 'Breakdown of element' column can have as many activities as you think is appropriate. Save the sheet as you work, as you will need to develop it for the next stage. When complete, the running total can be added together to give the total amount of time need to construct each individual element (example 1 to 25).

Note: This is different from the overall time taken to construct the house because there will be some elements that can occur concurrently, for example M&E (mechanical and electrical) first fixing can happen with plasterboarding, but on different floors at the same time. **This concurrent activity will be shown on the programme in a separate operation later in the process.**

Next stage

When the spreadsheet is completed with all the elements, breakdown of elements and number of days, another two further columns can be added that will capture the tolerances for TIME and COST. The construction industry refers to this as TIME OVER-RUN and COST OVER-RUN, both of which must be fully calculated and appreciated prior to the contractor submitting their programme of tender cost to the client.

Work breakdown: Construction of 24 Palace Avenue						
Project name	Element	Breakdown of element	No. of days	Tolerances adjustment description (RISK)	TIME OVER-RUN tolerance in days	COST OVER-RUN tolerance £
New house construction at 24 Palace Avenue	1 Drainage	1.1 Internal drainage	4	Groundwater may need to be pumped away	1	100.00
		1.2 External drainage	3	Groundwater may need to be pumped away	1	100.00
	2 Foundations	2.1 Internal foundations	5	Wet weather may delay the concrete pour	2	200.00
		2.1 External foundations	5	Wet weather may delay the concrete pour	2	200.00
	3 External walls	3.1 Blockwork	12.5	Wet weather may delay blockwork	3	300.00
		3.2 Brickwork	1.5	Wet weather may delay blockwork	3	300.00
	4 Floor slab	4.1 Prepare sub-base	1	Wet weather may delay preparing the slab	2	200.00
		4.2 Lay membrane	1	Wet weather may delay preparing the slab	2	200.00
		4.3 Pour concrete	1	Cold weather may delay pouring the slab	2	200.00
	5 Internal walls	5.1 Set out studs	4	Wet weather may delay fixing structural studs	1/2	50.00
		5.2 Fix sole plates	5	Wet weather may delay fixing structural studs	1/2	50.00
		5.3 Fix head restraints	5	Wet weather may delay fixing structural studs	1/2	50.00
		5.4 Fix vertical studs	3	Wet weather may delay fixing structural studs	1/2	50.00
		5.5 Fix noggin and pattress	3	Wet weather may delay fixing structural studs	1/2	50.00
	6 Roof structure					
	7 Roof structure					
	8 Roof coverings					

Before the cost is submitted to the client, the contractor's team, often led by the directors of the business, will conclude what total risk allowance will be added to both the programme (in days or weeks) and tender sum (in cost).

For example, the contractor will appreciate that if the building is proposed to be constructed in the middle of winter on a very exposed site with a high water table and restricted access to the site, then it is likely that they will increase the risk factor for the total number of days they will need to construct it and the total cost that the additional days will add onto the preliminaries.

They may also clarify or qualify within a document attached to their tender cost, clauses in the contract preventing the client from penalising the contractor if the inclement (poor) weather affects them and delays the handover.

This may be discussed at length between the client and contractor during the pre-award of the contract (known as negotiating) depending on the method of procurement and who the client is accountable to.

APPLICATION / Construct your own detailed spreadsheet for the work breakdown of your course and submit it to your teacher. Get used to working with software for making spreadsheet documents and look at online sources to support you with this. Remember, the objective is to calculate how long a quantified activity will take to construct and to understand how external factors such as weather, ability, skills shortages, size of project, location, time of year and so on can affect the total programme and cost outcome.

Finally, continue building your spreadsheet by adding a calendar at the top and highlight relevant cells below showing the durations in days or weeks. Show concurrent activities by overlapping them in a step-like sequence, as shown below.

Programme for Construction of 24 Palace Avenue

ELEMENT	Breakdown of element	No. of days including tolerances (RISK)	January				February				March				
			6th	13th	20th	27th	3rd	10th	17th	24th	2nd	9th	16th	23rd	30th
			2	3	4	5	6	7	8	9	10	11	12	13	14
1. Drainage	1.1 internal drainage	5	MTWTF												
	1.2 external drainage	4		MTWT											
2. Foundations	2.1 internal foundations	6		F	MTWTF										
	2.1 external foundations	6			TF	MTWT									
3. External walls	3.1 blockwork	13.5				WTF	MTWTF	MTWTF	M						
	3.2 brickwork	12.5					TF	MTWTF	MTWTF	M					
4. Floor slab	4.1 prepare sub-base	2							TF						
	4.2 lay membrane	2							TF						
	4.3 pour concrete	2								MT					
5. Internal walls	5.1 set-out studs	5								MTWTF					
	5.2 fix sole plates	6								TWTF	MT				
	5.3 fix head restraints	6								WTF	MT				
	5.4 fix vertical studs	4								TF	MT				
	5.5 fix noggin & pattress	4									MTWT				

Assessment Guidance

Introduction to the assessment process

In this section, you will discover how to understand the way that the assessments are marked and how to correctly write answers to the questions asked during it and when submitting your coursework. This will help you to answer the questions as thoroughly as possible and also help you to demonstrate your skills, knowledge, understanding and application of Constructing the Built Environment.

As with most types of assessment, the greater the accuracy and depth of your answer, the more accurate your answer will be, and the greater number of marks you should receive.

By reading and understanding this section you are preparing yourself for the assessment and improving your chances of success.

Overview of the of the assessment

The Constructing the Built Environment course has been designed for the 14- to 19-year-old group of learners, although this book will also be helpful to anyone who works within the industry. This is because the course is vocational and therefore many of the characteristics of the course can be applied to real work scenarios and situations.

Every unit has a purpose, which means that the outcomes are defined and can be either contextualised in your written or chosen response or, in some cases, realised. This is because you will be able to demonstrate the level of your knowledge and understanding by applying your skills to materials using the methods and techniques that you have learnt.

This holistic learning experience will allow you to deliver real tasks and may help you to discover new skills that you wish to develop further. In doing so, you may also discover that producing this type of work can be rewarding and realise further just how vital the construction industry is to both our economy and society, as well as the wider modern civilisation.

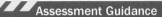
The composition of the units and how they are assessed can be seen in the following table, together with the guided learning hours that indicate the minimum amount of time required to be taught, instructed and to participate in the process.

WJEC Level 1/2 Vocational Award in Constructing the Built Environment			
Unit number	Unit title	Assessment	GLH (guided learning hours)
1	Safety and security in construction	External	30
2	Developing construction projects	Internal	60
3	Planning construction projects	External	30

External assessment – Units 1 and 3

e-Assessment

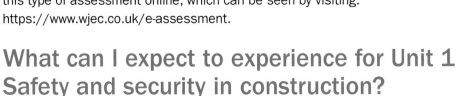

'e-Assessment', WJEC.

Units 1 and 3 will be assessed by 'e-assessment' and this involves the use of ICT. This gives you the opportunity to complete the assessment on-screen. Similarly, the marking will also be done in the same way, meaning that there is no need to use the traditional pen and paper. Other than this, there is no difference in content between on-screen and paper versions. The awarding body (WJEC) has a demonstration of this type of assessment online, which can be seen by visiting: https://www.wjec.co.uk/e-assessment.

What can I expect to experience for Unit 1 Safety and security in construction?

Unit 1 has a 60-minute on-screen examination with a total of 60 marks. There is also an audio-enabled version (available on request), so, if necessary, ensure that your centre is aware of the need for this when your course leader registers for the examination or at any time after registration.

The questions will be a mix of short and extended answer questions, based on stimulus material, which is a visual, verbal and/or auditory method used to communicate certain ideas to enable the examiners to stimulate your knowledge and understanding of security and safety in the built environment.

This means that every question will have an applied problem-solving scenario to help you visualise and understand the context of how to answer. In doing so, each paper will assess all learning outcomes (LOs).

Assessment criteria will be sampled in each series of questions, so that there is a reasonable selection of questions for you to demonstrate your knowledge and understanding.

The assessment will be available to take in the summer of each year, generally at the centre where you are learning the course. If a learner is unsuccessful, then then are allowed one re-sit opportunity.

Those who successfully achieve higher grades will see the highest grade contribute towards the overall grade for the qualification. Consequently, the awarding body has produced a mark scheme that will be used as the basis for marking the examination papers as follows:

- Graded Level 1 Pass
- Level 2 Pass
- Level 2 Merit
- Level 2 Distinction.

What can I expect to experience for Unit 3 Planning construction projects?

Unit 3 is the longer on-screen examination. The total time allowed for this is 120 minutes and, once again, has a total of 60 marks. There is also an audio-enabled version (available on request), so, if necessary, ensure that your centre is aware of the need for this when your course leader registers for the examination or at any time after registration.

The questions will be a mix of short and extended answer questions, based on stimulus material which is a visual, verbal and/or auditory method used to communicate certain ideas to enable them to stimulate your knowledge and understanding of security and safety in the built environment.

Every question will have an applied problem-solving scenario to help you visualise and understand the context of how to answer.

In doing so, each paper will assess all learning outcomes (LOs).

Assessment criteria will be sampled in each series of questions so that there is a reasonable selection of questions for you to demonstrate your knowledge and understanding.

The assessment will be available to take in the summer of each year, generally at the centre where you are learning the course. If a learner is unsuccessful following their results, then they are allowed one re-sit opportunity.

The grading of this unit will be awarded based on the following performance descriptions. These performance descriptions give a broad indication of the standards of achievement likely to have been shown by candidates awarded corresponding grades for external assessment.

This means that the grade awarded will depend upon the extent to which the candidate has met certain descriptors. Sometimes, any shortcomings in candidates' responses in some aspects of the examination may be counterbalanced by better responses in others.

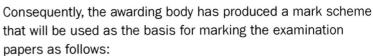

Consequently, the awarding body has produced a mark scheme that will be used as the basis for marking the examination papers as follows:

- Graded Level 1 Pass
- Level 2 Pass
- Level 2 Merit
- Level 2 Distinction.

Candidates must score at least a one Level 1 pass in each of the units to be successful in the qualification; see the following table.

Unit number	Points per unit			
	Level 1 Pass	Level 2 Pass	Level 2 Merit	Level 2 Distinction
1	1	4	5	6
2	2	8	10	12
3	1	7	5	6

Sample assessment material and past papers can be found on the WJEC website. These are freely available for you and your teacher to download and use as part of your revision and exam preparation. You will need to obtain both the question paper and the mark scheme (the answer sheet). You may need your teacher to explain the mark scheme, especially the answers to longer, high-mark questions.

You might find that some questions are difficult or that you don't know how to answer a question. This is a good indicator that you need to revise these areas to improve your knowledge. Don't be tempted to skip these questions for others that are easier. Filling in the gaps in your knowledge and understanding now is the best way to improve your marks in the real exam.

What will be externally assessed?

The assessment criteria for Units 1 and 3 form the route you should take to achieve the learning outcomes and it's likely that there will be a question relating to each of these in the assessment. The summary table opposite shows the assessment criteria. Check that you understand them all, prior to taking the assessment, by revising each one and logging in your notebook the progress you are making. Understand the percentage of each learning outcome in context of the assessment, for example in Unit 1 LO3 look how valuable the percentage of understanding 'minimise risks to health and safety' is to, potentially achieving up to 40% of the total marks available.

Unit 1 Safety and Security

Learning outcomes	Assessment criteria	Marks	%
LO1 Know health and safety legal requirements for working in the construction industry	AC1.1 Summarise responsibilities of health and safety legislation AC1.2 Identify safety signs used by construction industry AC1.3 Identify fire extinguishers used in different situations AC1.4 Describe role of the Health and Safety Executive	12–18	20–30
LO2 Understand risks to health and safety in different situations	AC2.1 Identify hazards to health and safety in different situations AC2.2 Describe potential effects of hazards in different situations AC2.3 Explain the risk of harm in different situations	12–18	20–30
LO3 Understand how to minimise risks to health and safety	AC3.1 Explain existing health and safety control measures in different situations AC3.2 Recommend health and safety control measures in different situations	18–24	30–40
LO4 Know how risks to security are minimised in construction	AC4.1 Identify risks to security in construction in different situations AC4.2 Describe measures used in construction to minimise risk to security	6–12	10–20
	TOTAL	60	100%

Unit 3: Planning Construction Projects

Learning outcomes	Assessment criteria	Marks	%
LO1 Know job roles nvolved in realising construction and built environment projects	AC1.1 Describe activities of those involved in construction projects AC1.2 Describe responsibilities of those involved in construction projects AC1.3 Describe outputs of those involved in realising construction projects		
LO2 Understand how built environment development projects are realised	AC2.1 Describe processes used in built environment development projects AC2.2 Calculate resources to meet requirements for built environment development projects AC2.3 Assess potential effect of factors on project success AC2.4 Interpret sources of information	19–25	32–42
LO3 Be able to plan built environment development projects	AC3.1 Sequence processes to be followed AC3.2 Apportion time to processes AC3.3 Set project tolerances	20–26	33–43
	TOTAL	60	100%

What does 'good' look like?

Fairly good: Level 1 Pass

Learners recall, select and communicate limited knowledge and understanding of the content of Units 1 and 3. They analyse and evaluate limited information to apply limited understanding to questions and tasks set. Learners present information with limited clarity.

Good: Level 2 Pass

Learners recall, select and communicate some knowledge and understanding of the content of Units 1 and 3. They analyse and evaluate some information to apply some relevant understanding to questions and tasks set. Learners use some effective written communication skills to present information that is mainly clear and accurate.

Really good: Level 2 Distinction

Learners recall, select and communicate detailed knowledge and through understanding of the content of Units 1 and 3. They analyse and evaluate information to apply relevant understanding to questions and tasks set. Learners use effective written communication skills to present information clearly and accurately.

(Source: 'WJEC Level 1/2 Award in Constructing the Built Environment: Specification (December 2018)', https://www.eduqas.co.uk/qualifications/constructing-the-built-environment/Constructing%20the%20Built%20Environment%20-%20Level%2012%20Award%20-%20Specification%20(from%202018).pdf)

Re-sitting

Learners may have one re-sit opportunity for each externally assessed unit. The higher of the two outcomes will contribute to the final qualification grade.

Command verbs and meanings

Command verbs are words used within the assessment by the examiner to express to the student what the examiner wants them to do. This is because assessment questions should test a range of knowledge and skills. These questions should test (by asking a relative question) and reward (by awarding relative marks) for critical appreciation and the ability to apply what has been learnt rather than reproducing facts that have been memorised.

Command verbs

Below are definitions of 'be able to' related command verbs, used in both the context of the assessment and within the construction industry.

Collaborate – make a real contribution to the work of a team, supporting team members as required.

For example: The team of bricklayers collaborated really well by sharing the loading out of materials and sharing the cleaning-up duties, as they constructed the brick panel safely and productively.

Communicate – ensure information is delivered and received effectively.

For example: The communication between the site manager, crane driver and gate person was really effective as they used short-wave radios and text messaging to control who entered and left the site during the time that the mobile crane was lifting. This meant that the lifting activity was safer because communication was excellent between the team members.

Display – organise and present information diagrammatically.

For example: The project manager displayed the organisational chart within the canteen so that visiting contractors could understand the roles and responsibilities of the contractor's management team. The job titles were organised, with Carys the project manager at the top of the diagram, then Dylan the site manager below her and so on.

Handle – manipulate a tool/equipment to a desired effect.

For example: Ananya the bricklayer handled the pointing trowel with great precision as she formed uniform recessed pointing in the fresh mortar of the brickwork panel.

Maintain – keep in an appropriate condition.

For example: The condition of the carpenter's chisels was first class – they were sharp and free of rust. They were clearly maintained to a high standard, which reflected his workmanship as well.

Monitor – observe and record activity, could also include ensuring expected progress is maintained.

For example: Jamil the safety officer observed the behaviour of the scaffolders at work. Jamil recorded the activities of the scaffolders who were working in the city centre as he monitored their performance for his behavioural record sheet.

Plan – organise a range of components into a logical sequence. This could include timings and also how the organisation is presented.

For example: The timber skirting boards and door linings were staked flat on the floor where they could not be tripped over or damaged. Sara, the labourer, was good at planning the logistics of the workplace.

Present – organise and communicate in a way that can be clearly followed and understood. Often refers to oral communication skills and may include use of supporting information.

For example: The site induction was well presented because the site manager organised the safety posters on the canteen wall in a sequence that told the story of her safety expectations of the workforce, working from left to right. This also meant that the operatives could read the same posters at break time while they had a coffee.

Process – use a series of actions to elicit results.

For example: The process for using burning gear on site started by visiting the site safety officer and requesting a hot work permit. This permit could only be issued if you presented him with a risk assessment. If a permit was issued then it would also have to be signed off by the site manager as part of the process.

Record – obtain and store data and information.

For example: Tim the site store person would obtain signatures from anyone who was issued with new PPE. This data would then be stored on the site PPE register and recorded to ensure PPE was not misused.

Use – employ something for a purpose.

For example: Rhiannon employed the new mechanical spray-plastering technique as she wanted to speed-up production and use something far more efficient for large areas of work.

Internal assessment – Unit 2 Developing construction projects

This unit will be assessed internally by the centre where you are registered for this course. This means that the unit will be assessed via a summative controlled assessment.

This is an evaluation by your teacher of what you have learnt by comparison to the standards expected of Level 1 and Level 2. This evaluation will only take place at the end of the instructional Unit 2. The centre must follow the guidelines published by the awarding body to ensure consistency.

To help with delivering the best and most consistent outcomes for Unit 2, a model assignment brief for Sunnycove Lodge provides the scenario for this and is available digitally on the awarding body's website:
https://www.eduqas.co.uk/qualifications/constructing-the-built-environment/ Constructing%20the%20Built%20Environment%20-%20Level%201-2%20Award%20 -%20Unit%202%20-%20iSAM.pdf?language_id=1.

'Sample Assessment Materials', WJEC.

The internal assessment is based on a PLAN, DO and REVIEW approach to learning, so the use of check sheets and mark sheets applies here. The mark sheet for this unit can also be downloaded from the awarding body's website (above) and this provides a transparent, consistent and straightforward check sheet for teachers and learners alike.

Performance bands are provided for Level 1 Pass, Level 2 Pass, Level 2 Merit and Level 2 Distinction. Evidence must clearly show how the learner has met the standard for the higher grades.

There are three stages of assessment that will be controlled:

1 *Task setting: What the task is and what to do.*

2 *Task taking: How to do the task and provide evidence.*

3 *Task marking: How to mark the task and check against the evidence provided.*

By interpreting technical information you will plan the refurbishment of Sunnycove Lodge (a 19th-century building) and take account of health and safety issues. Using the appropriate skills and techniques you will carry out the refurbishment.

This will involve interpreting technical information to identify materials, tools and equipment needed to complete construction tasks. This is your opportunity to develop a range of construction skills that can be used during construction processes, ensuring you take account of any health and safety issues. Download the check sheets provided in Unit 2 of this book to plan and check your works as you progress. These are vital in demonstrating compliance with both your ability to identify resources (Unit 3 AC2.1) required and to evaluate the quality of the tasks you deliver (Unit 3 AC3.3) as they develop and hopefully improve.

LINK

The check sheets can be seen on pages 112–113.

Task	What to do	AC	Specifically designed resources from this book
1	**Under supervision:** Working on your own, produce a work plan over a period of 2½ hours. You can use the following resources: • ICT software calculator • class notes • provided technical information. *No feedback can be given to you on your performance during this time.*	**AC1.1** Interpret technical sources of information **AC1.2** Plan sequence of work to meet requirements of sources of information **AC2.1** Identify resources required to complete construction tasks **AC2.2** Calculate materials required to complete construction tasks **AC2.3** Set success criteria for completion of construction tasks	**AC1.1** (pages 10–15) **AC1.2** (pages 16–19) **AC2.1** (pages 27–30) **AC2.2** (pages 31–34) **AC2.3** (pages 36–37)
2	**Under supervision:** Working on your own, produce observation records and photographs over a period of 11 hours. You can only use the following resources: • Class notes • Access to all materials and equipment required to complete the construction tasks, access to work plan produced in task 1 and technical information. *Only once your work is marked can you be given feedback on your performance.*	**AC2.4** Prepare for construction tasks **AC3.1** Apply techniques in completion of construction tasks **AC3.2** Apply health and safety practices in completion of construction tasks	**AC2.4** (pages 86–91) **AC3.1** (pages 92–105) **AC3.2** (pages 106–110)
3	**Under supervision:** Working on your own, produce a report with supporting photographs over a period of 1½ hours. You can only use the following resources: • class notes • ICT software • work plan. *Only once your work is marked can you be given feedback on your performance.*	**AC3.3** Evaluate quality of construction tasks	**AC3.3** (pages 111–112).

Glossary of Key Terms

106 agreements: Conditions imposed by the local authority planning department such as: for every 300 houses constructed the developer must provide a nursery school; or contribute to an adjacent road widening project; or maybe provide a pedestrian bridge over a local busy highway that the occupants of the new housing are likely to use. This ensures that any new development is safe and sustainable for people to live, work or play in and not subsidised by the tax payer.

4D: Four-dimensional; 4D information relates to data such as time (when and where details were formed or building components such as lighting, control panels or roof lights were installed) and programme (where and when proposed activities or series of linked activities such as steel work and its related floor construction is to be or has been formed). This allows the construction team to plan the construction process or the client and user to see details of the construction process after the building is completed. This can help them identify defects and resolve maintenance issues.

Abut: For example, where a building, part of a building or boundary of land meet with one another. 'The external north elevation of the farmhouse abutted the public footpath, so the owner decided that they wanted to use mirrored glass within the window frames to prevent pedestrians seeing into the kitchen.'

Aesthetic: In the construction industry, the term 'aesthetic' is used to describe the visually pleasing standards of a building, parts of building or overall impression of a particular detail of the building that has been formed. This pleasurable and descriptive term is often used when contrasting details such as: 'The black and grey natural Welsh slated conical roof turret contrasted with the green leaves of the trees behind the building. It was aesthetically pleasing to see the building in context of the adjacent valley.'

Attribute: Quality or feature.

Backfill: The process of replacing spoil materials into a trench or excavation, often using a mechanical device such as an excavator.

Bevel: A guide that can be altered to form a desired and consistent angle, often used when working with wood and metal.

BIM: Building information modelling; the process of managing, distributing and providing accessible storage of data relating to a building project. This data will be in the form of electronic models, drawings, schedules and manufacturers' information, which can be accessed, reviewed and revised during the lifecycle of the building or refurbishment.

Brownfield site: A site that has previously been developed.

Calibrated: The successful process of checking the measurement of an instrument.

CDM regulations: Legislation that promotes the project team to:

- sensibly plan the work so the risks involved are managed from start to finish
- have the right people for the right job at the right time
- cooperate and coordinate your work with others
- have the right information about the risks and how they are being managed
- communicate this information effectively to those who need to know
- consult and engage with workers about the risks and how they are being managed.

(Source: http://www.hse.gov.uk/construction/cdm/2015/index.htm)

China-graph pencil: A special pencil that can write on smooth surfaces as well as paper and remains water resistant when wet.

Collaborate: Willingly and positively help to create or contribute to a project.

Compliant: An acceptable level of pre-agreed standards.

Concurrent: Activities that can happen at the same time, for example bricklaying and roofing.

Construction project lifecycle: The length of time it will take to plan, construct and handover the building project.

COSHH: The law that requires employers to control substances that are hazardous to health.

CPD: Continuing professional development; a diverse range of courses and self- delivered, life-long learning that professionals often do to maintain a current level of understanding and knowledge of their relative role or job. This approach often helps them attain more formal qualifications and also develop new skills.

Crazing: The random and irregular pattern often created when cementitious materials (things made from cement) dry out and crack in an uncontrolled fashion.

CSCS: Construction Skills Certification Scheme; an accredited body that certifies construction qualifications.

Cupping: The natural, localised bending across the cross-section of timber usually as it dries out.

Cybercrime: Electronic crime using the internet.

Defective: Work or a system that does not work as it was designed to do or has relevant flaws within it and is therefore non-compliant.

Discharge: The successful process of ensuring an obligation such as a planning condition is achieved.

Dry-bond: The process of offering-up a material and placing it in-situ (in place) and assessing its suitability, prior to bonding it in place permanently.

Dutyholders: Those that plan or carry out the work.

Easements: The legal right to cross over, near or by another party's land or part of their infrastructure, like a gas or water main, for a reason.

Efflorescence: When fine salt-like chemicals migrate (move) to the surface of materials such as brickwork and render, often using rainwater as a vehicle. The result of this migration is white powder residue that is left behind as the drying-out process occurs. In severe cases, solid white salt-like deposits become crusty and solidify, which are very unsightly and difficult to remove.

Eliminate: Get rid of entirely.

Employee: Someone who works under an employment contract. A person may be an employee in employment law but have a different status for tax purposes. Employers must work out each employee's status in both employment law and tax law.

Employer: A person or organisation who employs people under an employment contract.

Exemplary: An example of the best of its kind.

Expansion/cracking joints: Joints intentionally formed in concrete to control the location and extent of the natural tendency of concrete material to crack.

Flammable: The ease by which something is set alight and the rate at which it burns.

Float: Spare/extra time that may be needed.

Foul sewers: The network of foul (human waste) drainage that exists in the UK to transport human waste to treatment works where the waste is removed and the water treated.

Future proofing: How the building can adapt and cope with change in the future.

Government Soft Landings (GSL): A building delivery process which runs through the project, from inception to completion and beyond, to ensure all decisions made during the project are based on improving operational performance of the building and meeting the client's expectations. (Source: BSRIA, 2018, https://www.bsria.co.uk/services/design/soft-landings/)

Graticules: A defined set of consistent horizontal and/or vertical lines that help define measurement.

Hold-point: A point during the design or construction process where the team pauses, checks and agrees the condition, progress or quality of an important aspect of the task at hand such as a vulnerable detail, an important part of a process or inspecting a building component that is about to be covered-up by the next activity.

Housekeeping: The approach to day-to-day operational cleanliness and storage.

Identify: Recognise, distinguish and establish what something is.

Impression: In the construction industry, clients, through their design team, often refer to the word impression to relay a concept or feeling of a building or space within the built environment. For example, 'The castellated stonework gave the impression of a castle wall and so made the building's envelope feel robust and more secure.'

Infrastructure: Important building and transportation networks.

Intellectual property: includes concepts, designs and literature which are protected by copyrights, trademarks and patents to ensure that business branding and unique selling points are kept as trade secrets and not stolen.

Interfaces: Where and how things relate to one and other.

Irrational storage: Storage that could cause an accident or or allow the materials being stored to be damaged.

Lifecycle of the building: The period of time from when the building is completed to the end of its useful life, which usually is followed with its demolition and recycling.

Likelihood: The chance something will happen or the probability of something becoming a reality.

Milestone dates: Dates in the programme or schedule that reflect important points in the design process and construction phase such as when the concept design is anticipated as being completed or when the building becomes structurally completed, watertight or ready for handover.

Mitigate: Get rid of as much as possible.

Modular: Sometimes referred to as volumetric construction, or prefabricated buildings, although, strictly speaking, a prefabricated building need not be modular. Buildings made up from components manufactured on assembly lines in factories then assembled on site in a variety of arrangements. (Source: https://www.designingbuildings.co.uk/wiki/Modular_buildings)

Monitor: To observe and conclude progress on a task or project.

Off-cut: The waste material left over from the process of cutting a material such as wood or plasterboard.

Organised crime groups: Criminals who target specific vulnerable victims.

Parameters: requirements or measurable aspiration.

Perp: relates to the perpendicular mortar joint that is present on all masonry. This is the shorter vertical joint that is often finished in different ways depending on the specification.

PPE: Personal protective equipment can include gloves, goggles and hard hats.

Prefabrication: A manufacturing process that promotes the construction of building components prior to their installation on site.

Quantification: The end product of a process to conclude the gross (before wages, tax and other expenditure) and net (after paying tax) sum of a particular task or a group of tasks.

RAL: A colour-matching system, mainly used for varnishes and powder coatings, but also for plastics.

RAMS: Risk assessment and methods statement; term used in the construction industry referring to two key documents that are needed to be approved by the site manager prior to starting work.

Reactive guarding: The tactic of responding to the activation of a monitored automated alarm within a set timeframe. The response can be by a mobile operative attending the site but is more likely to be an audible warning demanding potential criminal and unwanted visitors to leave.

RIBA: Royal Institute of British Architects; one of the professional organisations that represents the architectural profession.

Risk: The likelihood that a person may be harmed if they are exposed to a hazard.

Severity: The degree of harm that could occur or a measure of how bad an injury could become.

SEWSCAP 3: South East and Mid Wales Collaborative Construction Framework; an organisation where several local authorities and other parties form a cooperative to improve their buying power for goods and services, and to ensure best value is obtained when public money is used to construct building and infrastructure such as schools, public buildings and roads.

Signed-off: A point at which the team agrees that the work constructed has reached an acceptable level of workmanship and complies with the specification. This is often recorded by the responsible person such as a site manager on a procedural form or electronically by email or digital acceptance.

Site induction: The formal presentation given to all new visitors and members of the workforce when they arrive on site and always before they enter the site on a visit or to work. A record of this safety critical presentation is kept to demonstrate that all stakeholders are aware of project-specific risks, hazards and objectives, as well as rules and restrictions.

Stakeholder: A member of a group with an interest in project.

Statutory application: The formal process of submitting an official procedural form or adhering to a recognised process.

Substrate: The material that is protected by and located below the surface material.

Sustainabilty: Within the construction industry relates to adequate appreciation, management and use of the limited three precious things that we all have in common:

1 Health: For example, the mental/physical health of an individual, commercial health of a business or the wellbeing of both.
2 Time: For example, programme, time constraints and limited or restricted opportunity to access the works.
3 Resources: The availability of materials, labour, plant, minerals, water, education and skills.

(All of the above are non-exhaustive.)

Template paper: a prepared document that can be used as an aid when delivering a task.

Tendering: The process by which bids are invited from interested contractors to carry out specific packages of construction work. It should adopt and observe the key values of fairness, clarity, simplicity and accountability, as well as reinforce the idea that the apportionment of risk to the party best placed to assess and manage it is fundamental to the success of a project.

(Source: https://www.thenbs.com/knowledge/tendering-for-constructionprojects)

Validate: Prove the accuracy.

VOC: Volatile organic compound. Substances, such as formaldehyde, that used to be in certain man-made materials, e.g. MDF or some paints, that are dangerous to health and are now restricted in use.

WAH: Working at height; relates to any activity where someone is engaged in a task that has a risk of falling (from any height) that could result in harm.

Workability: The capability of a product when it is worked.

Index

Photo Acknowledgements

All photos in this book have kindly been supplied by ISG plc, except for the following.

p.1 Ant Clausen; p.6 ISG WJEC Construction / YouTube; p.10 (top) maziarz; p.10 (bottom) stockfour; p.11 Pormezz; p.12 Jirsak; p.13 (top) Studio DMM Photography, Designs & Art; p.13 (middle) Brian A Jackson; p.13 (bottom) Sarawut Aiemsinsuk; p.14 (top) Gorodenkoff; p.14 (bottom) Sirtravelalot; p.15 (top to bottom) hartphotography, Michael Grot, Aunging, andre quinou; p.19 Gustavb / Creative Commons Attribution-Share Alike 3.0 Unported license; p.16 mikolajn; p.17 (top to bottom) Safety Signs and Notices, Arcady, P Studio, Viktorija Reuta, Standard Studio, Ogdesign, Safety Signs and Notices, MiMieKo, Walther S; p.18 (1) Walther S, p.18 (2) Thomas Pajot, (3) Thomas Pajot, (4) Thomas Pajot, (5) Trish Volt, (6) PATIWIT HONGSANG, (7) Thomas Pajot, (8) Walther S, (9) Adisak Panongram, (10) Thomas Pajot, (11) Nevada31, (12) canbedone, (13) Thomas Pajot, (14) Thomas Pajot, (15) Trish Volt, (16) Walther S, (17) pockygallery, (18) Thomas Pajot, (19) Walther S, (20) Thomas Pajot, (21) Arcady, (22) Walther S, (23) Arcady, (24) Adisak Panongram, (25) Arcady, (26) Thomas Pajot, (27) Thomas Pajot, (28) Adisak Panongram, (29) PATIWIT HONGSANG, (30) Adisak Panongram, (31) Shutterstock, (32) Thomas Pajot, (33) Thomas Pajot, (34) Thomas Pajot, (35) Thomas Pajot, (36) Thomas Pajot, (37) Arcady, (38) Walther S, (39) Trish Volt, (40) hanohiki, (41) Thomas Pajot, (42) hanohiki, (43) Thomas Pajot, (44) Trish Volt; p.20 (top left) Mark William Richardson, p.20 (top right) Viktorija Reuta, p.23 (bottom left) Craig Dingle; p.23 (bottom middle) Petrunovskyi; p.24 (top) Courtesy HSENI; p.25 (middle) Contains public sector information published by the Health and Safety Executive and licensed under the Open Government Licence; p.26 bongkarn pelthntod; p.27 Photographee.eu; p.28 Opsorman; p.30 (top left) zuphatra; p.30 (top right) Kandapa; p.30 (middle left) Aggapom Poomitud; p.30 (middle right) komkrit Preechachanwate; p.30 (bottom left) propae; p.30 (bottom right) Voyagerix; p.32 (top to bottom) Alfazet Chronicles, VendeDesign, Pixel Embargo, Alex_Murphy; p.34 (top to bottom) Alfazet Chronicles, VendeDesign, Pixel Embargo, Alex_Murphy; p.35 (top to bottom) MADSOLAR, Vectors Market, Classica2, BARS Graphics, bsd, Siberian Photographer, popular.vector; p.38 (top) daseaford, p.38 (bottom) Rawpixel.com; p.39 (top left to right) SIM VA, Redline Vector, VectorA, garagestock, Sunshineart; p.39 (bottom) alphaspirit; p.40 (clockwise) Redline Vector, SIM VA, Sunshineart, VectorA, garagestock; p.41 (top left to right) Gagnar, lumar_art, BagirovVasif, nanovector; p.42 (top clockwise) Redline Vector, SIM VA, Sunshineart, VectorA, garagestock; p.42 (top left to right) Gagnar, lumar_art, nanovector, BagirovVasif, p.42 (bottom four) VendeDesign, Shutterstsock Vector, VectorA, Blan-k; p.43 (1) jpreat, (2) lenetstan, (3) 3dphoto, (4) Rido, (5) Dragance137, (5) Fotofermer, (6) Iakov Filimonov, (7) Dragance137, (8) Mikael Damkier, (9) wavebreakmedia, (10) Fotofermer, (10) Dragance137, (11) richardjohnson, (12) tomaso79; p.43 (centre) leolintang; p.44 (top) pixinoo; p.44 (bottom) GlebSStock; p.45 spacezerocom; p.47 richardjohnson; p.48 (left to right) Vector Forever, Artco, richardjohnson, richardjohnson; p.49 (top) SergeBertasiusPhotography; p.49 (middle four) VendeDesign; p.49 (bottom) hans engbers; p.50; p.51 Tasha and Deki; p.52 Natasa Adzic; p.53 Tashatuvango; p.54 (top) Dragon Images; p.54 (middle) Sarawut Aiemsinsuk; p.54 (bottom) Worawee Meepian; p.56 Wright Studio; p.57 (top to bottom) NicoElNino, AmalPhoto, Artem Stepanov, rolandtopor, Romolo Tavani; p.58 (top left to right) Tashatuvango, Suwatchai Pluemruetai, Dragon Images, Sarawut Aiemsinsuk, Cineberg; p.58 (bottom left to right) AmalPhoto, Artem Stepanov, rolandtopor; p.59 (top) AmalPhoto; p.59 (bottom) Romolo Tavani; p.60 (top) Dmytro Prikhodko, p.60 (bottom) oatawa; p.61 (top) everything possible; p.61 (bottom) Freedomz; p.62 (top) Artem Stepanov; p.62 (bottom) rolandtopor; p.63 PopTika; p.64 (top) Gaf_Lila; p.64 (middle) VictorN; p.64 (bottom) WindAwake; p.65 (top) Martin Designer; p.65 (middle) Lotus_design; p.65 (bottom) DaryaSuperman; p.66 (top) Francescomoufotografo; p.66 (middle) Marcel Derweduwen; p.66 (bottom) Matthew Dickson; p.68 (top) urfin; p.68 (bottom) Mizin Roman; p.69 Virojt Changyencham; p.76 ilkecelik; p.78 (top) Okcm; p.78 (bottom) Christos Theologou; p.79 (all top left to right)

Strejman; p.79 (bottom clockwise) Hein Nouwens; MADSOLAR; AmalPhoto; Junjun123; Artem Stepanov; AmalPhoto, white snow; Vector Icon Flat; p.80 (top to bottom) Hein Nouwens, MADSOLAR, AmalPhoto, Junjun123, AmalPhoto, AmalPhoto; p.81 (top to bottom) AmalPhoto, white snow, Vector Icon Flat, Artem Stepanov; p.82 Photo Smile; p.83 (top to bottom) Africa Studio, Vasileios Karafillidis, bogdanhoda, martinho Smart; p.84 (top to bottom) RomanR, Bilanol, eggeeg, Spok83, Radomir Rezny; p.85 (top) SVIATLANA SHEINA; p.85 (bottom) xstock; p.86 VendeDesign;p.87 Anar Babayev; p.107 (top) brizmaker; p.108 (bottom) Alex Farias; p.109 (top) Barry Barnes; p.109 (bottom) Martin Mehes; p.110 Teguh Mujiono; p.111 (top) Syda Productions; p.111 (middle top to bottom) AmalPhoto, Artem Stepanov, rolandtopor; p.111 (bottom) Bannafarsai_Stock; p.112 (top to bottom) Picons Me, p.114 (bottom) creatIR76; p.116 (top) Olena Hromova; p.117 Georgejmclittle; p.118 michaeljhung; p.120 Roman023_photography; p.121 KomootP; p.122 Syda Productions; p.123 StockCube; p.124 (top) Rob Byron; p.124 (bottom) mmohock; p.125 oneinchpunch; p.126 (top) Tyler Olson; p.126 (middle) 1Roman Makedonsky; p.126 (bottom) DB; p.129 (top) Christian Delbert; p.130 Monkey Business Images; p.131 (bottom) Lucky Business; p.145 (top) Aleksandar Tasevski; p.131 (top) King Ropes Access; p.132 (top) Dmitry Kalinovsky; p.133 (bottom) goodluz; p.135 (top) MIND AND I; p.135 (bottom) Shine Nucha; p.136 (top) ALPA PROD; p.136 (bottom) GaudiLab; p.137 (top) Romeo Pj; p.137 (middle) SeventyFour; p.137 (bottom) Morakot Kawinchan; p.138 (top) kyozstorage_stock; p.138 (middle) Iryna Liveoak; p.138 (bottom) OH HO; p.138 (2 up) Ian Francis; p.144 (middle) mrmohock; p.145 (top) Aleksander Tasevyski; p.145 (bottom) Alexander Lysenko; p.147 (bottom) Mykhubh; p.148 hxdyl; p.149 Federico Rostagn; p.154 (top) Ovidiu Dugulan; p.154 (bottom) Pixel-Shot; p.155 Dmitrii Iarusov; p.156 Grand Warszawski; p.157 emzet70; p.158 Photo wind; p.159 Andy Dean Photography; p.161 Iakov Filimonov; p.166 Courtesy FSC; p.167 (bottom) Courtesy BSRIA p.173 (top) fizkes; p.173 (bottom) IVASHstudio; p.174 DesignPrax; p.175 (top) keport; p.175 (bottom) fizkes; p.176 (top) IVASHstudio; p.176 (middle) fizkes; p.176 (bottom) IVASHstudio; p.178 (top) Antonio Guilliem; p.178 (bottom) Jacob Lund; p.179 Bannafarsai_Stock; p.180 (top) Oleksii Didok; p.180 (bottom) Koy_Hipster